Principles of telecommunication-traffic engineering

D. BEAR, B.Sc., F.S.S., F.I.M.A.
The General Electric Company Limited
Hirst Research Centre
England

PETER PEREGRINUS LTD.
on behalf of the
Institution of Electrical Engineers

Published by: The Institution of Electrical Engineers,
London and New York

Peter Peregrinus Ltd., Stevenage, UK, and New York

First published 1976
© 1976: Institution of Electrical Engineers
Second impression 1978

Published as paperback edition 1980
© 1980: Institution of Electrical Engineers

Bear, Donald
    Principles of telecommunication - traffic
    engineering. - (Institution of Electrical
    Engineers. IEE telecommunication series;
    2).
    1. Telephone systems
    I. Title   II. Series
    621.385'1   TK6401   80-40410

ISBN: 0 906048 36 2

Typeset at the Alden Press, Oxford, London and Northampton
Printed in England by A. Wheaton & Co., Ltd., Exeter

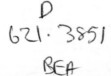

Preface

The telecommunication industry is undergoing rapid change owing to the introduction of new techniques, including replacement of electromechanical by electronic
switching, and the control of switching by computer-type processors. In the
development and application of new systems, economic utilisation of plant is no
less important than the design of equipment and circuits. A system must be capable
of handling peak loads efficiently, at a cost compatible with the provision of
service to subscribers on acceptable terms. An appreciation of the principles of
traffic engineering is therefore essential to engineers concerned with the development
and planning of telecommunication switching systems. It is hoped that this monograph will help to supply such a need, and will also be useful as a basic introduction
for students and intending 'teletraffic' specialists (this convenient term has long
replaced the more cumbersome, and perhaps too narrow, phrase 'the application of
the theory of probability in telephone engineering and administration').

The monograph is concerned with general principles rather than their application
to particular systems, which are only discussed for the purpose of illustration. The
amount of attention paid to the Strowger system may, perhaps, seem high in view
of its approaching obsolescence; the simple trunking of this system, however,
makes it particularly suitable for illustrating traffic principles, while the advantages
of more flexible modern systems are most easily understood in the context of its
limitations. Extensive traffic tables and other reference data are not included, as the
monograph is not intended to be a working manual of traffic-engineering practice.
They will, however, be found in the list of References, especially nos. 1, 7, 9, 20, 21
23, 25, 80, 92, 98, 103, 108, 110, 142, 149, 150 and 153.

The mathematical level has been kept as elementary as the aim of clearly
expounding the basic principles of the subject allows. Thus, in the Chapter on
waiting-call systems, only the case of Poisson input, negative exponential holding-
time distribution and first-in-first-out service is analysed in any detail. References
to specialist literature are given for the benefit of readers wishing to follow up more
advanced topics.

Many practical traffic problems are too complex for exact mathematical solution,
and approximation is necessary. There are often a considerable number of alternative
approximate solutions to a particular problem; it is impossible to deal with all of
them in the compass of this volume, so that many valuable contributions to the
subject have inevitably been ignored.

The previous literature of telecommunication traffic includes comprehensive
textbooks by G.S. Berkeley[50] and R. Syski,[51] to both of which the author is
heavily indebted. The former, however, is out of date, while the latter is more
advanced mathematically than the present work. A few other books dealing with

limited aspects of the subject are included in the references (e.g. nos. 17, 22, 52—56 and 155), along with a number of specialist articles. In general, preference has been given to those dealing with practical methods of general application rather than highly theoretical work, or discussions of the traffic problems of particular systems.

The author wishes to acknowledge the benefit he has derived over many years from discussion and correspondence with teletraffic specialists and others in the telecommunication field, both in the United Kingdom and overseas: including his colleagues with the Hirst Research Centre and the Telecommunications Division of The General Electric Company, Limited, and former colleagues with the Telecommunications Division of Associated Electrical Industries, Limited. The monograph owes more to the experience of the British Post Office in traffic engineering than may be apparent from explicit References, and the author is indebted to J.A. Povey, A.C. Cole and other members of Telecommunications Headquarters staff for access to much useful information.

Most of the problems in Appendix 2 were originally prepared as tutorial and examination questions for a postgraduate course in Telecommunications Technology at the University of Aston in Birmingham. The author is grateful to Prof. J.E. Flood for inviting him to lecture on the subject, and hopes that the monograph may have benefited therefrom.

Contents

Glossary

All technical terms are explained in the text. For convenience, however, definitions of some of the most frequently used terms are given in this glossary. Most of them are quoted from the report of the Nomenclature Committee to the 5th International Teletraffic Congress, New York, 1967 (preprints of technical papers, pp.588–592).

Alternative routing (Alternate routing in the USA)
A procedure whereby several routes are searched to complete a connection. The different routes involve different switching stages or switching networks.

Availability (Accessibility)
The number of appropriate outlets in a switching network which can be reached from an inlet.

Blocking (see Congestion)

Busy hour
The uninterrupted period of 60 min for which the traffic is a maximum.

Call congestion (ratio)
The ratio of the number of call attempts which cannot be served immediately to the number of call attempts offered.

Common trunk
A trunk accessible from all groups of a grading.

Conditional-selection system (see Link system)

Congestion
The condition where the immediate establishment of a new connection is impossible owing to the unavailability of paths.

Congestion function
Any function used to relate the degree of congestion to the traffic intensity; e.g. Erlang's loss function $E_{1,N}(A)$.

Crosspoint
Each connection in a switching network is established by closing one or more crosspoints. A crosspoint comprises a set of contacts that operate together and extend the speech and signal wires of the connection.

Delay system
A switching system in which a call attempt, which occurs when all accessible paths for the required connection are busy, is permitted to wait until such a path becomes available.

Effective (or equivalent) availability
For a conditional-selection system, the availability of a single-stage reference switching network which, with the same number of trunks, handles the same traffic with the same probability of congestion.

Full availability
Full availability exists when any free inlet can reach any free outlet of the desired route

regardless of the state of the system.

Grade of service
Any practical interpretation of a congestion function.

Grading (or graded multiple)
A grading is obtained by partial commoning or multipling of the outlets of connecting networks when each network only provides limited availability to the outgoing group of trunks.

Grading group
A unit within a grading in which all inlets have access to the same outlets.

Holding time
The total duration of one occupation. There are many classes of holding time, e.g. setting-up, conversation etc.

Homogeneous grading
A grading in which the outlets of an identical number of grading groups are connected to each outgoing trunk.

Individual trunk
A trunk which serves only one group of a grading.

Internal blocking or internal congestion.
In a conditional-selection system, the condition in which a connection cannot be made between a given inlet and any suitable free outlet owing to the unavailability of paths. (Note: the term 'matching loss' is used in the USA).

Limited availability
Exists when access is given only to a limited number of trunks of a given route.

Link
The device in a switching network connecting one outlet of one of the connecting stages to an inlet of the next connecting stage. The term is used mainly in conditional-selection systems.

Link system or conditional-selection system
A system in which: (i) there are at least two connecting stages (ii) a connection is made over one or more links (iii) the links are chosen in a single logical operation (iv) links are seized only when they can be used in a connection.

Loss system or lost-call system
A switching system in which call attempts fail when there is no free path for the required connection.

Mean occupancy (or load per device)
The traffic carried by a group of devices divided by the number of devices in the group.

Occupation
Each use of a device regardless of cause.

O'Dell grading
A progressive grading in which only identically numbered outlets of adjacent groups are commoned.

Overflow traffic
When facilities are provided to handle the traffic that cannot be carried by a group of circuits, the traffic which is not carried by the group in question is called overflow traffic.

Own-exchange call
A call between two subscribers connected to the same exchange.

Partial common trunk
A trunk accessible from more than one but not all groups of grading.

Progressive grading
A grading in which the outlets of different grading groups are connected together in such a way that the number of grading groups connected to each outgoing trunk is larger for later choice outlets.

Skipping
The interconnection of identically numbered choices of nonadjacent grading groups.

Slipping
The interconnection of differently numbered choices of grading groups.

Time congestion (ratio)
The ratio of the time for which congestion exists to the total time considered.

Traffic flow or traffic intensity (of traffic carried)
The traffic volume occurring during a specified period of time divided by the duration of the period, both quantities being expressed in the same time unit. The unit of traffic intensity is the erlang.

Traffic offered
A calculating quantity having the dimension of traffic intensity. It is equal to the mean number of call attempts occurring during the mean holding time. (Note: In the USA, the terms 'offered load' and 'carried load' are used synonymously with 'traffic offered' and 'traffic carried', respectively.)

Traffic volume
The sum of the holding times of a certain number of occupations.

Transposed multiple
A form of skipped grading used particularly for subscriber concentration stages in which a subscriber shares some outlets with one group of subscribers and other outlets with other groups of subscribers.

Trunk
Usually equivalent to 'link', but not restricted to conditional-selection systems.

Waiting call system (see Delay system)

Chapter 1

Outline of telecommunication switching

Like any other public service, a telecommunication system has to cater for a fluctuating demand which can only be predicted within a limited degree of accuracy. The nature of the service requires a high standard of performance; it will have failed in its objective unless the great majority of demands are satisfied with little or no delay. At the same time, transmission and switching equipment is expensive and must be efficiently utilised. Optimisation of the network structure and the provision of equipment is therefore one of the most important aspects of telecommunication engineering.

1.1 Historical background

The first practical telephone was invented by A.G. Bell in 1876. The first manual exchange was installed in 1878 (New Haven, Connecticut, USA), and the first automatic exchange in 1892 (La Porte, Indiana, USA). In the early days of telephony, each administration developed its own methods of determining the provision of equipment and the standard of service, usually on an empirical basis. With the increasing complexity of telephone networks and the introduction of automatic switching, however, it became clear that a more scientific approach was necessary to ensure a proper balance between service and cost. Because of the random nature of telephone traffic, this involved the use of probability theory. Indeed, the early telephone-traffic engineers were pioneers in applied probability, and were responsible for the development of techniques which were much later applied to operational research in other fields, such as road traffic and production control.

The first mathematical analysis of telephone traffic was probably an unpublished memorandum by G.T. Blood (USA) in 1898.[51] During the first half of the twentieth century, traffic theory was actively developed by a small number of investigators in different countries, some of whose contributions are mentioned later. At first, many telephone engineers regarded the subject as abstruse and of doubtful value, but it soon established itself as an indispensible aid to the design and planning of switching systems.

The rapid development of telecommunication technology since 1950, including the introduction of electronic techniques, posed many new traffic problems. This led to increased interest in the subject on the part of telecommunication administrations, manufacturing industry, and universities. The first International Teletraffic Congress was held in Copenhagen in 1955, and subsequent congresses have been held triennially in different countries. Summaries of the proceedings of most of the congresses have appeared in technical journals of the host countries.[46-49, 60]

1.2 Single-stage switching

Fig. 1 shows a simple switching network serving a self-contained 10-line telephone exchange. It comprises a matrix with ten inlets and ten outlets represented by horizontal and vertical lines, respectively. Each telephone line is connected to one inlet and one outlet; any inlet can be connected to any outlet by means of a device, the 'crosspoint', located at their intersection. One such connection is shown in Fig. 1a, the crosspoint in operation being encircled. Only one crosspoint in each row,

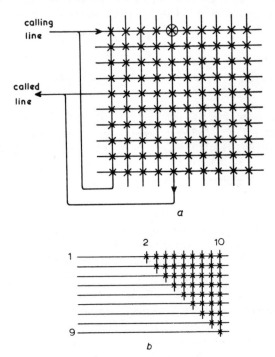

Fig. 1 Single-stage switching network

and one in each column, can be in use simultaneously. Since it is not necessary to connect a subscriber to his own line, the main diagonal crosspoints are not used. The arrows indicate the path of a call; they represent the direction in which the call is set up, not the direction of speech. In traffic theory, it is not usually necessary to distinguish between forward and return transmission paths, so that both can be represented by a single line. Nor is it necessary to specify the physical means of connection. The crosspoints may be electromechanical or electronic devices, and the transmission paths may be separate physical conductors, or, in a multiplexed system, separate carrier frequencies or time slots using a common conducting path.

In this simple example, it would be possible to reduce the number of crosspoints to 45, while still enabling any line to call any other line (Fig. 1*b*). The saving would be offset by circuit complication, since the mode of connection would depend on the inlet and outlet concerned. Moreover, this type of switch does not lend itself so readily to the construction of more complex systems than that of Fig. 1*a*.

1.3 Multistage switching

The scheme of Fig. 1*a* is usually uneconomical, because the number of crosspoints, which account for a substantial proportion of the switching costs, increases as the square of the number of lines. It can be reduced by replacing the simple matrix by a series of matrices.

Fig. 2 shows a typical 3-stage switching scheme for a 1000-line exchange; for simplicity, the crosspoints are not shown. Each telephone is connected to an inlet of a switch in the A-stage and an outlet of a switch in the C-stage. The lines

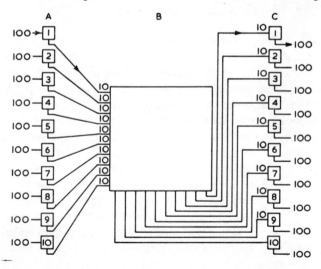

Fig. 2 Step-by-step trunking

connecting switches A and C to B represent ten speech channels each. The first stage (A) concentrates the traffic from 100 inlets to 10 outlets; thus, blocking of calls may occur through all the outlets of a switch being occupied, unlike the non-blocking arrangement of Fig. 1. It is possible to construct nonblocking multistage networks (Chapter 8). In practice, however, blocking usually has to be tolerated to reduce equipment costs to an economical level while maintaining acceptable standards of service (Section 2.8).

The terms 'blocking' and 'congestion' are equivalent; they refer to the condition where a connection cannot be immediately completed due to the unavailability of a free path.

The second stage (B) switches the traffic to the required route, according to the called number. The third stage (C) completes the connection and reverses the traffic concentration. The number of crosspoints in each switch is the product of the numbers of inlets and outlets, the total being

$$10 \times 1000 + 10\,000 + 10 \times 1000 \ = \ 30\,000$$

as compared with 1 000 000 for the single-stage scheme of Fig. 1*a*.

In the simple scheme of Fig. 2, switch rank A has no routing function, but merely concentrates the traffic for the purpose of economy. On the other hand, switch rank B directs the traffic to the required routes without altering its concentration, the number of inlets and outlets being the same. In practice, however, the distinction is not always as rigid; a routing switch may also alter the traffic concentration.

1.4 Conditional selection

Fig. 2 is typical of a 'step-by-step' system, in which each stage is controlled independently. It is possible, however, to control several stages together, in such a way that the seizure of a link is prevented unless it can form part of a through connection. Systems with this feature are known as 'conditional-selection' or 'link' systems. This enables smaller switches to be used, as shown in Fig. 3, which requires only 21 000 crosspoints to serve 1000 lines. This scheme is unsuitable for step-by-step switching because, on seizing an A-switch outlet, a call would have access to only one C-switch inlet, which might have a high chance of being busy. With conditional selection, however, the A-switch outlet would not be seized unless the associated C-switch inlet were free. This example is intended only to illustrate the principle; it is not suggested that Figs. 2 and 3 are precisely equivalent in traffic capacity, or that they are necessarily optimal networks of their kind.

A link system may comprise any number of stages, and the number of possible connection patterns is very large. The best pattern depends on the cost of crosspoints and other elements, and on various constraints arising from standardisation of switch sizes, simplification of control etc. The simple patterns shown in Figs. 4*a*

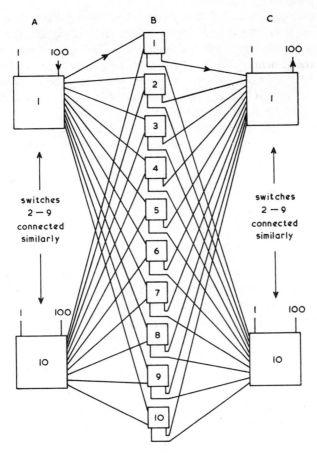

Fig. 3 Link (conditional-selection) system trunking

and 5*a* will serve to illustrate some of the features of practical link systems. For the sake of clarity, the switch sizes are smaller than would normally be the case in practice, and the switches are shown in simpler form than before. In both cases there are 30 subscribers' lines, connected to the inlets of the A-switches and the outlets of the D-switches. The lines are divided into two groups of 15, denoted I and II. In Fig. 4*a*, each line has access to 3 A–B links, and then to six B–C links, three to each subscriber group. The C and D switches are connected in the same manner as the B and A switches. Traffic concentration from 18 A–B links to 12 C–D links may be justified by relative link costs and other considerations.

For the purpose of traffic calculation, it is often convenient to represent only the paths available to a particular call or class of calls. Let us consider a call between any two subscribers in group I. The available paths are shown in Fig. 4*b* in the form of a network graph. It will be seen that they form a series–parallel network. This pattern is particularly simple to analyse in terms of traffic theory, and also tends to

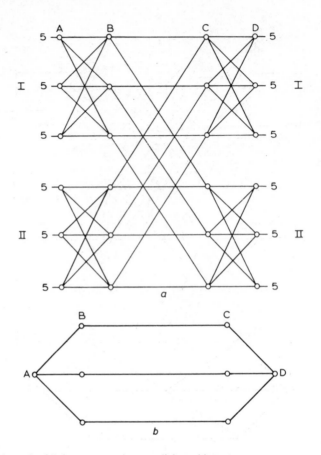

Fig. 4 Network graph of link system: series—parallel trunking

simplify the equipment and control arrangements. The network of Fig. 3 is also of this type. It is not always, however, the most efficient arrangement.

In Fig. 5*a* the B and C switches are square, so there is no traffic concentration between links. Figs. 5*b* and 5*c* show network graphs for calls between the same and different groups, respectively. It will be seen that the former have five and the latter four alternative paths. Such a lack of symmetry is generally undesirable, but cannot always be conveniently avoided. This does not arise in the case of Fig. 4*b*, which applies to both classes of call.

Greater efficiency can be obtained by arranging for the links to carry traffic in either direction, at the cost of circuit complication. This is usually easier with a link than a step-by-step system, the latter being directionally oriented. Fig. 6 shows a 1000-line exchange with conditional selection and bothway links. Each line is connected to the exchange at one point only. The total number of cross-points is the same as in Fig. 3, but the number of separate switches is reduced.

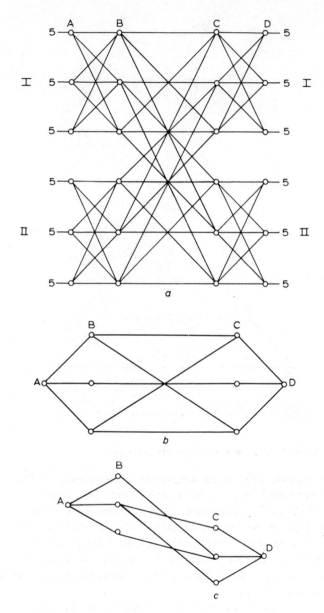

Fig. 5 Network graph of link system

Moreover, a call between different A switches has a choice of 20 routes instead of 10, so there is less chance of blocking.

There may be more than one link between a pair of switches. Thus, in Fig. 3, the

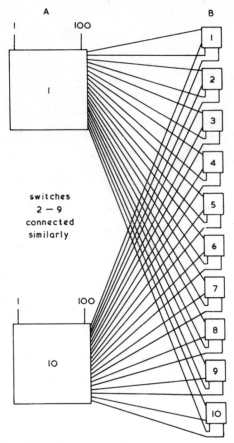

Fig. 6 Link system with combined inlets and outlets

number of B switches could be reduced from 10 to 5, each having two links to each A switch and to each C switch. This would increase the traffic capacity at the cost of a considerable increase of crosspoints, and possibly complication of control.

The term 'link', in the sense of a device connecting an outlet of one stage in a switching network to an inlet of the next, is mainly, but not exclusively, applied to conditional-selection systems. In traffic theory, the term 'trunk' is generally used in the same sense, without being confined to link systems or to long distance (trunk) circuits.

1.5 Combined step-by-step and conditional selection

It is possible to apply conditional selection to a number of consecutive stages in a step-by-step system, so that a call does not enter any of these stages unless there is a free path through them, but takes its chance of blocking at a subsequent stage

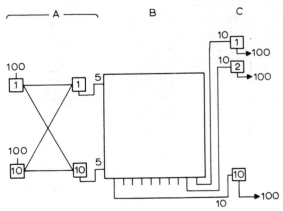

Fig. 7 Combined step-by-step and link system

which is not included in the conditional selection. An example is shown in Fig. 7. A 2-stage link system concentrates traffic from the subscribers' lines on to the first routing switch, before dialling; a link is not seized unless it gives access to a free routing-switch inlet. The subsequent switches form a step-by-step system, controlled by the dialled digits. This scheme enables every call to have access to every routing-switch inlet instead of only ten as in Fig. 2. As a result, the number of inlets can be reduced, in this particular case, by 50%, so there is a substantial saving to offset the cost of the additional switching rank.

Similar schemes were used in a number of older telephone systems; with the techniques available, it was easier and cheaper to control two predialling stages as a link system than the more complex dial-controlled stages.

1.6 Common control

In some systems, the devices which operate the crosspoints are associated with individual inlets or outlets of the switching matrix, and are held for the whole duration of the call. In others, a single control may serve a number of inlets or outlets, and is released after use. This slightly complicates the circuit and increases the blocking, since a free link may be briefly unavailable when the associated control is in use. As a result, a few more links may have to be provided, but the cost is offset by the saving in relatively expensive control equipment.

The simplest form of common control is illustrated in Fig. 8, in which a set of four controls operates a row of crosspoints in each of three matrices. Since an inlet cannot be connected to an outlet unless the appropriate control is free, the links and controls together form a conditional-selection system.

Fig. 9 shows an alternative scheme in which a common pool of three controls serves all twelve links. If a link is free, it is unavailable only if all controls are in use, so it is generally possible to make do with fewer controls than in the scheme of Fig. 8. On the other hand, the cost of the extra crosspoints must be set against this saving.

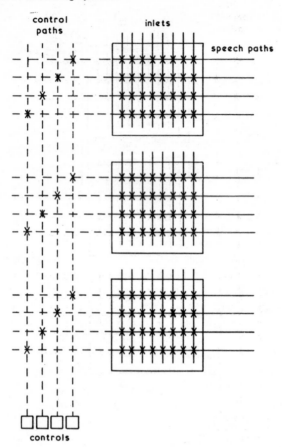

Fig. 8 Simple common-control scheme

The next step is to adapt the control to serve a number of switching stages together. This reduces the number of controls required and facilitates the operation of conditional selection, although it is not essential; conditional selection can be effected by means of signals between separately controlled stages. Most modern systems, however, are of the common-control type; the use of complex common controls has been facilitated by advances in electronic techniques.

The term 'control' covers a number of operations involved in establishing a connection, other than the actual transmission of speech or data. It may be advantageous to allocate some of these to separate functional units. A typical control complex for a telephone exchange with electromechanical switches and electronic controls might comprise the following units, *inter alia*:

(a) *registers*; to store the number dialled by the caller until the connection has been established. Their traffic load depends on the average dialling time, which might

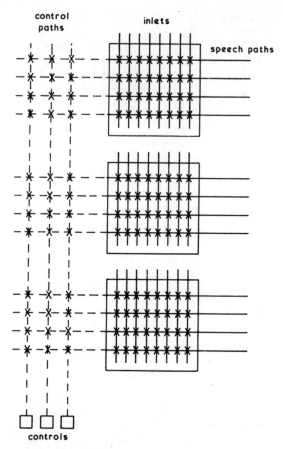

control paths

inlets

speech paths

controls

Fig. 9 Fully-available common-control scheme

be, say, a tenth of the average call duration, so that a substantial number may be required in a large exchange.

(b) *route control*; to determine the route of the call in accordance with the dialled number and the paths available. This operation can be extremely fast, so that a single control may be sufficient.

(c) *markers*; to operate the crosspoints as instructed by the route control. With mechanical crosspoints, the setting time is significantly longer than the purely electronic operation of the route control, so that several markers may be required, each serving a section of the exchange.

These functions are shown diagrammatically in Fig. 10.

The first common-control device in the history of telephony was, of course, the human operator. The relevance of the work of the early traffic theorists to manual operation was not fully appreciated at the time, with the result that manual-exchange

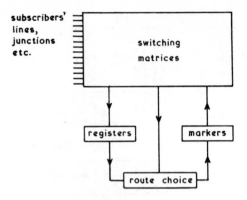

Fig. 10 Elements of typical common-control exchange

traffic practice evolved independently on an empirical basis. Later on, however, the increasing integration of manual and automatic switching in telephone networks demanded a unified approach to traffic problems, and this is now generally accepted.

1.7 Alternative routing

In Fig. 3, it would be possible to eliminate the intermediate switch-rank, B, by providing sufficient direct links between the A and C switches. As the traffic between each pair of switches is relatively small, however, this would require a disproportionately high number of links (Chapter 3). A compromise is to provide a small number of direct links as first choice; calls which cannot find a free direct link are routed via a secondary switch. Thus, in Fig. 11, each A-switch has one direct link to each C-switch and two links via stage B. Provided there is sufficient traffic to keep the direct links highly occupied, the scheme results in a saving in intermediate crosspoints with little loss of efficiency.

A somewhat similar principle is employed in re-entrant (entraide) trunking, as shown in Fig. 12. Connection is made of one or more outlets in each A-switch to the other A-switch. Thus, if it is required to connect an inlet of A_1 to an outlet of B_1, but the link between them is occupied, the call may be routed via A_2, so that blocking is reduced at the cost of extra crosspoints. In more complex arrangements, an outlet of one stage may be connected to an inlet of a previous stage.

1.8 Switching techniques

Whilst this monograph is not primarily concerned with switching technology, it would not be complete without a brief account of the subject, since the evolution

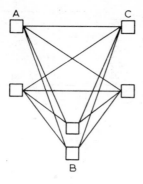

Fig. 11 Exchange trunking with alternative paths

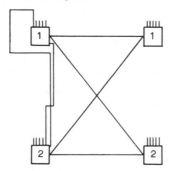

Fig. 12 Re-entrant (entraide) trunking

of telecommunication-traffic practice has been determined by the possibilities and limitations of the techniques available.

The development of large automatic exchanges was originally made possible by means of the 2-motion selector invented by Almon B. Strowger in 1889. In its simplest form, this comprises ten arcs of ten outlets, stacked vertically, as in Fig. 13*a*; for simplicity, only one contact per outlet is shown. The inlet is connected to a set of wipers mounted on a shaft, which moves vertically and then rotates to make connection with any outlet. Another type of switch, the uniselector, rotates in one plane only (Fig. 13*b*).

One possible method of employing these switches in the scheme of Fig. 2 is as follows (Fig. 14): each row of crosspoints in stage A represents ten outlets of a uniselector, so that each matrix contains 100 uniselectors; the inlet of a uniselector is connected to a subscriber's line. Corresponding outlets of all uniselectors are connected together, forming the ten columns of the matrix, each of which is connected to the inlet of a 2-motion selector. Thus, each row of crosspoints in stage B represents 100 outlets of a 2-motion selector, divided into ten levels. Corresponding outlets of all the selectors are connected together, forming the 100 columns of the matrix (ten per level). Each level is connected to the inlets of ten 100-outlet

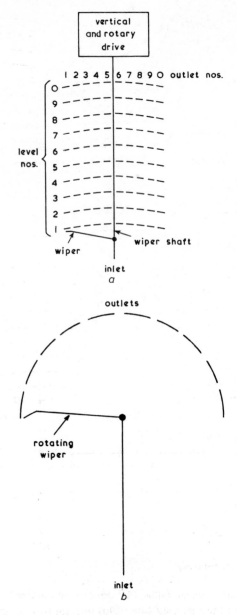

Fig. 13 Elements of Strowger switches

2-motion selectors, which are represented by one matrix in stage C. Corresponding outlets of these are connected together, and then to 100 subscribers' lines. Fig. 14 shows one A matrix and one C matrix, with links to the B matrix. When a caller

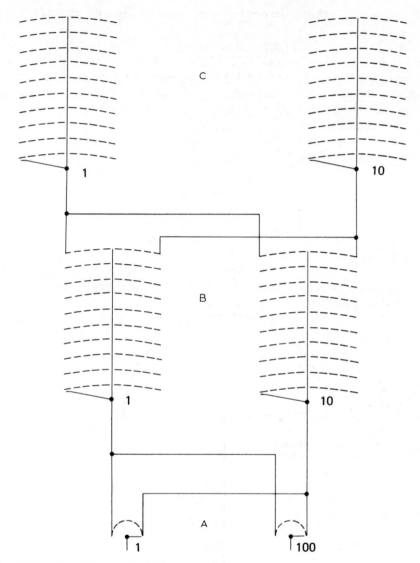

Fig. 14 Principle of Strowger trunking

lifts his receiver, his uniselector (stage A) searches for a free 2-motion selector (stage B). The dialling of the first digit causes this selector to step to the corresponding level, where it rotates to find a free outlet to a final selector (stage C). The dialling of the second and third digits causes the final selector to move vertically and horizontally to connect with the required line.

The 1000-line exchange contains 1000 uniselectors, which are relatively cheap, 100 final selectors, and 100 intermediate or group selectors; so called because they select a group of 100 lines, leaving the final selector to reach the particular called line.

The uniselectors can be replaced by 2-motion selectors used in reverse, as 'line-finders'. In this scheme, each matrix in stage A represents ten 100-outlet selectors. Corresponding outlets of these are connected together and thence to the subscribers' lines; thus the *inlets* of the matrix represent the *outlets* of 2-motion switches, and vice versa. When a caller lifts his receiver, a common-control device (not shown) allocates one of the linefinders, which immediately makes connection to the calling line by vertical and rotary motions. The line is thereby connected to a group selector (stage B) which responds to the dialled impulses as before.

Other systems employ large uniselectors instead of 2-motion selectors. The outlets are divided into groups corresponding to the levels of a 2-motion selector. They are faster than the latter, and the outlet grouping is more flexible.

Strowger switches, and other types of switch that are directly controlled by dialled impulses, are inherently more suitable for step-by-step than for conditional selection. Moreover, the individual crosspoints are as a rule relatively cheap; the most expensive part of the switch is the selector mechanism, which is associated with a row or column of crosspoints (selector outlets) depending on how the switch is used. Switches with large outlet capacities are therefore economical, so long as the mechanism is fast enough to make connection during dialling. Reduction of crosspoints is less advantageous than it is in the case of modern systems with relatively expensive crosspoint devices. Although, as already noted, some early systems incorporated conditional selection, the principle was not extensively applied until about 1950, when common-control systems based on the crossbar switch were developed. This consists essentially of a rectangular array of contact spring sets operated by horizontal and vertical bars, so that, unlike the Strowger switch, there is an obvious correspondence between its physical layout and the basic connection pattern of a switching matrix (Fig. 1). Other modern systems employ relays or, in the case of t.d.m. systems, electronic gates as crosspoints.

1.9 Interconnection of exchanges

For simplicity, the exchanges shown in Figs. 2—6 are assumed to be self-contained, whereas, in practice, it is usually necessary to accommodate junctions to and from other automatic exchanges and manual board services. In principle, it would be possible to use the same basic trunking scheme for a single exchange or a network of interconnected exchanges with a common numbering scheme. On the other hand, the relatively high cost of interexchange links imposes practical constraints on trunking flexibility. For example, it is generally inefficient to provide a number of self-contained, lightly loaded junction groups between a pair of exchanges; it is better to combine the traffic on to a common group of junctions and split it up at the distant end, even at the cost of extra switching equipment.

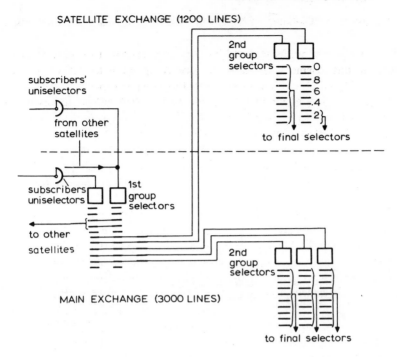

SATELLITE EXCHANGE (1200 LINES)

subscribers' uniselectors

from other satellites

2nd group selectors

to final selectors

subscribers uniselectors

1st group selectors

to other satellites

2nd group selectors

MAIN EXCHANGE (3000 LINES)

to final selectors

Fig. 15 Strowger satellite exchange scheme

To illustrate the influence of external routes on exchange trunking it is instructive to consider a local network of Strowger exchanges. Typically, this might comprise a central main exchange and a number of smaller 'satellite' exchanges. Figs. 15 and 16 show two possible trunking schemes. The conventional method of indicating a group of 2-motion selectors is by a square over ten horizontal lines; the square represents the connecting mechanism and relays, and each line represents the outlets of a level. The number of selectors in each group depends on the traffic, and is not shown. Each subscriber's line is connected to a uniselector.

The main exchange serves 3000 lines, and there are three satellite exchanges serving 500, 1000 and 1200 lines; the diagrams show the trunking of the main exchange and the largest of the satellites. For simplicity, only the routes of calls originating and terminating within this network are shown. Clearly, a minimum of four digits is required to identify all the numbers in the network. The routing of a call therefore requires at least three stages of 2-motion selection, including the 2-digit final selector. In Fig. 15 each first selector level gives access to up to 1000 lines, so it is necessary to reserve four first selector levels in the main exchange for outgoing junctions to the satellites; one level for each of the two smaller ones and and two levels for the larger one (Fig. 15). As already remarked, however, it is inefficient to divide the junctions on a route between two levels; moreover, the use of two 1000-line levels to serve only 1200 lines is wasteful, and reduces the number

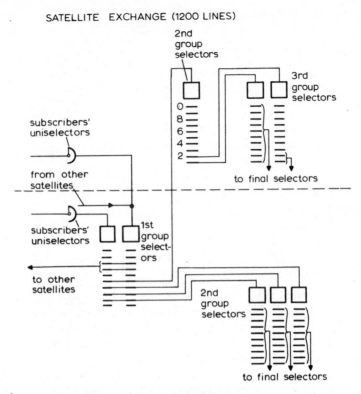

SATELLITE EXCHANGE (1200 LINES)

MAIN EXCHANGE (3000 LINES)

Fig. 16 Strowger satellite exchange scheme with mixed numbering

of spare levels available for future growth of the area. An alternative arrangement is shown in Fig. 16. This involves an extra rank of selectors in the large satellite exchange, so that five digits are now required to route a call there.

In Figs. 15 and 16 all satellite traffic, including calls between lines connected to the same satellite, is routed via the main exchange. Thus, the outlets of the satellite subscribers' uniselectors are connected direct to the first selectors in the main exchange. Unless the own exchange traffic is very small, however, this 'full-satellite' trunking may be wasteful of equipment. An alternative scheme is the discriminating satellite, as shown in Fig. 17. The satellite uniselectors are connected to 2-motion selectors in their own exchange. These selectors are known as discriminating selector repeaters (d.s.r.), and each one is associated with another uniselector, the junction finder. When a caller lifts his receiver, his uniselector seizes a d.s.r., and its junction finder seizes a first selector at the main exchange. When the first digit is dialled, the d.s.r. and the first selector step in unison. If the digit indicates that the called line is connected to the same satellite as the calling line, the d.s.r. absorbs the digit (5 in this example) and returns to normal. When the next digit is dialled,

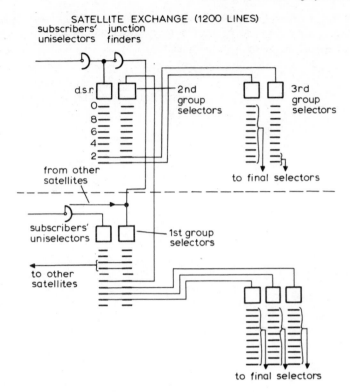

SATELLITE EXCHANGE (1200 LINES)

MAIN EXCHANGE (3000 LINES)

Fig. 17 Strowger discriminating satellite-exchange scheme

say, 2, the d.s.r. steps to level 2 and seizes a third selector, while the junction and first selector at the main exchange are released. The next two dialled digits operate the third and final selectors in the usual way. If the first digit is anything other than 5, the call is completed via the main exchange.

The need to adopt an awkward mixed numbering scheme to achieve trunking economy is an example of the restrictions imposed when the routing pattern closely corresponds to the numbering scheme. In large and complex networks, such restrictions can produce serious inefficiency. This can be avoided, even in a step-by-step system, by feeding the impulses, before the call is set up, into a device which translates them into the code required to set up the route, and transmits the appropriate impulses to operate the selectors. Thus, the number of digits dialled can be uniform although the number required to set up the call may depend on its route. This principle is employed in the Director system of the British Post Office.

In systems which employ end-to-end conditional selection within an exchange, external junctions can be connected at any convenient point. From the point of view of control simplification, it may be advantageous to connect the junctions in the same way as subscribers' lines, so that all calls are routed in a similar manner. Alternatively, they may be connected to intermediate switches, to reduce the number of crosspoints used by junction calls.

Chapter 2

Nature of telecommunication traffic

2.1 Traffic flow: the erlang

Let us consider a group of parallel connecting devices forming part of a communication network; e.g. the ten outlets of an A-switch, or the 100 outlets of all the A-switches shown in Fig. 2. The *traffic volume* carried by the group during a specified period is the sum of the times for which the devices are held. A more important concept is the *traffic flow* or *intensity*. This is defined as the traffic volume divided by the duration of the specified period; being the ratio of two time periods, it is a dimensionless quantity. The unit of traffic flow is named the *erlang* (abbreviated E) in honour of the Danish pioneer traffic theorist, Agner Krarup Erlang (1878–1929).[38] By definition, a single device, occupied continuously or intermittently for a total time t during a period T, carries t/T erlang, and its maximum possible load is 1 E. Thus, the number of erlangs carried cannot exceed the number of devices.

Let t_x denote the sum of the times during which exactly x out of N devices are held simultaneously, within a period T. Then

$$\sum_{x=0}^{N} t_x = T$$

The sum of the holding times of all devices is

$$\sum_{x=1}^{N} x t_x$$

Hence the traffic flow in erlangs is

$$\sum_{x=1}^{N} x t_x / T = \sum_{x=1}^{N} x(t_x/T)$$

Since t_x/T is the proportion of time for which x devices are held simultaneously, the right hand side of this equation is by definition the average number of devices held simultaneously during the specified period. This alternative definition of traffic flow is useful in traffice measurement.

If T is so long that the effect of unexpired calls at the beginning and end of the period is negligible, the traffic flow is approximately

$$(CT)h/T = Ch \text{ erlang}$$

where C = average number of device-occupations per unit time;
$\quad h$ = average holding time per occupation.
Hence the traffic flow is approximately equal to the average number of occupations occuring during a period equal to one average holding time. For practical purposes, this relationship can be regarded as exact (Section 2.2) and provides a third definition of traffic flow.

The terms 'occupation' and 'holding time' are used here in preference to 'call' and 'call duration' for the sake of generality; a device may be occupied for other purposes than the transmission of speech, and may be held for only part of the total call duration.

The above definitions apply to traffic actually carried. As the number of devices may be insufficient to meet all demands, it is necessary to distinguish between offered and carried traffic. In view of the third definition of traffic flow, it is natural to define traffic offered, in erlangs, as the average number of attempted occupations during a period equal to the average holding time of a successful occupation. The meaning of 'successful' depends on the function of the device under consideration. Thus, a successful occupation of a common control device may lead to an unsucessful attempt to set up a speech path.

Lost traffic is the difference between offered and carried traffic.

Needless to say, the term 'traffic flow' must not be taken to imply any analogy with the flow of fluids or vehicular traffic. It is true that a telephone call has a direction from calling to called subscriber, and, in this sense, it is meaningful to refer to traffic entering and leaving a switching point. Moreover, in the case of a step-by-step system, the setting up of a call involves progression through the network. These characteristics of the traffic process, however, are no part of the definition of traffic flow.

In American practice, traffic flow is sometimes expressed in terms of 'hundred call seconds per hour' for which the abbreviation CCS is used; i.e. the product of the average number of calls per hour and the average holding time, expressed as a multiple of 100 s.

Thus

$$1 \text{ erlang} = 36 \text{ CCS}$$

These basic definitions of traffic flow are applicable to any type of traffic, including speech and data transmission, operations involved in setting up a call, holding of circuits while a caller listens to a tone signal, false traffic due to faults etc. In the case of a data-transmission system, the traffic flow depends on the rate of arrival of messages (n), their average length (l), expressed in any suitable units such as characters or bits, and the rate of transmission (c characters, bits etc. per unit time). The traffic flow in erlangs is nl/c, the average holding time per message being l/c.

2.2 Traffic as a random process

So far, we have considered the actual traffic carried by a group of devices during a specified period. The occurrence of any particular pattern of call arrivals and terminations is obviously a matter of chance; it is, in effect, a sample from an indefinitely large population of possible traffic patterns.

More precisely, the number of devices out of N, occupied simultaneously at time t, may be denoted by a random variable X, which can take any integral value x in the range $0 \leqslant x \leqslant N$, with probability $p(x,t)$. It may be helpful to interpret this probability in terms of relative frequency (see Appendix 1.2). Let us postulate an indefinitely large number of identical groups of devices, all carrying the same average traffic flow, but independent of each other in respect of random traffic variations. Then $p(x,t)$ is the proportion of groups with x simultaneous calls at time t.

The traffic can be defined in the same way as before, but in terms of average instead of particular occupation patterns. The average number of occupied devices at time t is

$$\bar{X}(t) = \sum_{x=1}^{N} xp(x, t)$$

The expected traffic flow during a period T is obtained by averaging $\bar{X}(t)$ over the range $0 \leqslant t \leqslant T$, i.e.

$$A = \int_{t=0}^{T} \bar{X}(t)dt/T \text{ erlang}$$

As before, this expression also gives the average number of simultaneously held devices during the period T.

In many applications of traffic theory, it is assumed that a state of statistical equilibrium exists. This means that, while the number of occupied devices may fluctuate randomly, the chance of finding any specified number occupied is the same whenever the system is examined, so that

$$p(x, t) = p(x)$$

is independent of t.

The total proportion of time during which x devices are held simultaneously is approximately $p(x)$; the longer the state of equilibrium persists, the closer is the approximation. Thus, the instantaneous state of the group of devices can be regarded as a sample of its states during an indefinitely long equilibrium process, as well as a sample of the states of an indefinitely large collection of similar groups at one instant. These alternative sampling concepts lead to alternative definitions of equilibrium traffic flow

(a)   the *space* or *ensemble* average, obtained by averaging the number of occupied devices over a large number of similar groups at one instant

(b)   the *time average*, obtained by averaging the number of occupied devices over time in one group only.

In the case of equilibrium processes, these two averages, in general, have the same limiting value. Random processes, whether in equilibrium or not, which have this property, are called *ergodic*.

We have already seen that the traffic flow in erlangs, averaged over a long period, is approximately equal to the average number of occupations occurring during an interval equal to one average holding time, since the effect of unexpired calls at the beginning and end of the period is negligible. In a state of equilibrium, the equality of space and time averages implies that this relationship holds exactly even for a short period; since the former is the same at all points of the process, the latter must be independent of the duration of the period under consideration. This applies to the true average values of the underlying random variable, not necessarily to sample averages derived from any particular realisation of the process.

2.3 Traffic variations: the busy hour

The traffic in a particular public exchange normally falls to a low level at night, grows to a peak during the morning business period, slackens off at lunch time, and reaches another peak during the afternoon. There may be an evening peak as well due to social calls taking advantage of cheap rates.

These peaks are not as a rule very sharply defined; in fact, there is likely to be a period of at least an hour during which the traffic levels off, and fluctuates randomly without any general upward or downward trend, so that a state of statistical equilibrium exists. The uninterrupted period of 60 min during which the traffic is a maximum is known as the busy hour, and is generally used as the basis for traffic calculations. Shorter peaks can occur in special circumstances, such as breaks in popular television programmes, new year greetings etc. but it would be uneconomic to cater for all such eventualities.

The terms 'exchange busy hour' and 'group busy hour' are used to distinguish between periods of maximum traffic in the whole exchange, and in a particular group of trunks, which may or may not coincide. It is possible, although in practice unlikely, for the exchange busy hour to differ from the busy hours on all separate routes. For example, suppose the total originated traffic in an exchange is 80 E, 100 E, and 90 E during the first, second and third of three consecutive hours; while the traffic to others exchanges during these hours is 25 E, 50 E, and 55 E, respectively, the remainder being terminated at the same exchange. Clearly, the busy hours are the second for the total originated traffic, the third for outward traffic, and the first for own-exchange traffic.

The average busy hour traffic may vary on different days, so that successive daily busy hours cannot in general, be regarded as part of a continuous equilibrium process. These variations are of three kinds

(*a*)   long-term growth or decline of traffic
(*b*)   cyclical variations, weekly or seasonal
(*c*)   random variations, due to unpredictable factors affecting the general level of demand in an exchange or route on a particular day.

These must be distinguished from short random fluctuations in a state of equilibrium within a busy hour, which are due to chance irregularities in the pattern of call arrivals and terminations for a given average level of demand.

2.4 Pure chance (Poisson) traffic

The average busy hour traffic on a subscriber's line, including originated and received calls, is typically of the order of 0·1 E so that an average line is occupied for only about 10% of the busy hour. Hence about 90% of subscribers' lines are normally free during the busy hour, and, in a large exchange, this proportion is not likely to fluctuate very much. Moreover, it is reasonable to regard the subscribers as being relatively independent sources of traffic. While small groups of callers may tend to co-ordinate their calls, as with bookmakers' customers before a race, the effect is masked by traffic from a much greater number of unrelated sources.

It follows that the probability of a call arrival is approximately constant, irrespective of the state of the traffic, since, it is proportional to the number of sources which are not already occupied, and is unaffected by the arrival or non-arrival of other calls.

For theoretical purposes, these conditions which are only approximately satisfied in the real world, are assumed to hold precisely, unless the number of calling sources is relatively small (Section 3.5); in effect, they define a convenient mathematical model which is sufficiently realistic for traffic engineering. They are sometimes expressed by saying that the instants when calls arrive are distributed 'individually and collectively at random'. In other words, if time is represented by a straight line, and call arrivals by points, each point is placed at random independently of other points, and the number of points in any interval T is independent of those in any interval not wholly or partly included in T. This implies that the total number of points is not fixed in advance. Another term used to describe this type of process is 'pure chance' traffic. The random arrival of calls in this sense leads to a state of statistical equilibrium after a relatively short time; such a state can, however, result from many other arrival patterns.

In this context the term 'call arrival' means the first attempt to connect some device for the purpose of establishing a call. It is regarded as an instantaneous event. This is a legitimate approximation since the process of seizing a device and thereby rendering it unavailable to other calls is generally very short compared with its holding time after seizure.

Let us suppose such a process to continue for a very long time, and imagine this divided into very short intervals of duration dt. By making dt sufficiently short, we can ensure that the chance of more than one arrival is negligible. Let A denote the average number of arrivals per unit time; if the average holding time is adopted as the unit of time. A is the traffic offered in erlangs. A unit of time contains $1/dt$ intervals, so the probability that an interval, chosen at random, contains a call arrival is

$$\frac{A}{1/dt} = A\,dt$$

The probability that exactly x calls occur during a specified time T is the probability that exactly x out of T/dt intervals contain call arrivals, dt being chosen so that T/dt is an integer. Then x has a binomial distribution (Appendix 1.4), so that the required probability is

$$p_x = \frac{\left(\frac{T}{dt}\right)\left(\frac{T}{dt}-1\right)\left(\frac{T}{dt}-2\right)\ldots\left(\frac{T}{dt}-x+1\right)(1-A\,dt)^{(T/dt)-x}(A/dt)^x}{x!}$$

$$= \frac{T(T-dt)\,(T-2dt)\ldots(T-[x-1]dt)\,(1-A\,dt)^{-x}\{(1-A\,dt)^{1/dt}\}^T A^x}{x!}$$

as $dt \to 0$, $(1-A\,dt)^{1/dt} \to e^{-A}$
Thus

$$\dot{p}_x = \frac{(AT)^x}{x!}\,e^{-AT}$$

This is the Poisson distribution.
It will be shown later, by the method of equations of state, that this distribution applies to the number of calls in progress simultaneously, given unlimited traffic capacity, as well as to the number of call arrivals (Section 3.1).

Clearly, the mean value of x is AT. The variance of x is (Appendix 1, Section 6) (Mean value of x^2) $- (AT)^2$

$$= \{\text{Mean value of } x(x-1)\} + AT - (AT)^2$$

$$= \sum_{x=0}^{\infty} x(x-1)p_x + AT - (AT)^2$$

$$= \sum_{x=2}^{\infty} x(x-1)p_x + AT - (AT)^2$$

$$= (AT)^2 \sum_{x=2}^{\infty} e^{-AT} \frac{(AT)^{x-2}}{(x-2)!} + AT - (AT)^2$$

Putting $y = x - 2$, this becomes

$$(AT)^2 \sum_{y=0}^{\infty} p_y + AT - (AT)^2 = AT$$

2.5 Negative exponential distribution

Let a random instant be chosen during a Poisson process, not necessarily coinciding with a call origin. The distribution of the interval until the next call origin can be derived as follows:

As before, let the time axis be divided into short intervals dt. The time until the next call origin exceeds a specified value t if, and only if, the first, second ... (t/dt)th. intervals contain no call origins; the probability of this is

$$(1 - A dt)^{t/dt}$$

which tends to e^{-At} as dt tends to zero.

Hence the distribution function of t, i.e. the probability that the time until the next call origin is less than or equal to t, is

$$F(t) = 1 - e^{-At}$$

The probability-density function is $f(t) = \dfrac{dF(t)}{dt} = Ae^{-At}$ This is the negative exponential distribution. The mean value of t is

$$\bar{t} = \int_{t=0}^{\infty} tAe^{-At} dt$$

Substituting $u = At$, this becomes

$$\bar{t} = 1/A \int_{u=0}^{\infty} ue^{-u} du = 1!/A = 1/A$$

which is the same as the mean interval between call arrivals. This is to be expected, since the form of $F(t)$ is the same whether or not the random instant coincides with a call origin. It may, however, seem paradoxical that, when an interval is divided at random, the unelapsed part has the same distribution as the whole interval. It might be argued that, since the dividing point is equally likely to fall anywhere in the interval, the average value of the unelapsed part should be half that of the whole, i.e. $1/2A$, not $1/A$. The fallacy in this objection is that, roughly speaking, longer intervals occupy a higher proportion of total time than shorter ones, so that a random point is likely to fall in an interval of above average duration.

We have shown that Poisson distribution of arrivals implies negative exponential distribution of interarrival times. The converse is also true; assuming negative exponential distribution of interarrival times, the probability that a call occurs within a short interval dt after a random instant is

$$1 - e^{-Adt} = 1 - 1 + Adt \pm \text{ terms in } (Adt)^2 \text{ and higher powers} \rightarrow Adt \text{ as } dt \rightarrow 0$$

Hence the probability that a short interval contains a call arrival is independent of time and of previous call arrivals, which as we have seen, implies that the arrival times have a Poisson distribution.

The mean value of t^2 is

$$\int_{t=0}^{\infty} t^2 A e^{-At} dt$$

Substituting $y = At$, this becomes

$$1/A^2 \int_{y=0}^{\infty} y^2 e^{-y} dy = 2!/A^2 = 2/A^2$$

The variance of t is

$$\text{var}(t) = (2/A^2) - (1/A)^2 = 1/A^2$$

There is evidence that call durations often tend to have a negative exponential distribution, which implies that the future duration of a call, at any time during its progress, is independent of its past duration, since as we have seen, the part has the same distribution as the whole. The explanation of this apparently paradoxical feature of subscribers' behaviour is not altogether clear; it may reflect an element of randomness in the decision to terminate a telephone conversation.

In data communication systems, the message length is not continuous since it is made up of discrete units of information (bits, characters etc.). The geometric distribution is sometimes applicable. This means that the probability of the message continuing for one more unit, irrespective of its previous length, has a constant value p; hence the probability of a message length exceeding x units is p^x. In other words, the geometric distribution shares the property of 'lack of memory' with the negative exponential distribution, and is in fact closely approximated by the latter for long messages.

2.6 Traffic circulation in a network

In a step-by-step system, each switching stage has a slightly longer holding time than the next, depending on the dialling and connection times, and carries correspondingly higher traffic. Thus, in Fig. 2, a link between stages A and B is seized almost immediately a caller lifts his receiver, whereas a link between B and C is not seized until he has dialled the first digit. Allowing for dilatory callers the average interval before dialling commences may be, say, 4 s, while the average dialling time per digit may be, say, 1·5 s. If the average call-holding time from commencement of ringing to the end of conversation is 100 s, the average holding time of links A–B and B–C are 108·5 s and 103 s, respectively. The total traffic on the A–B links therefore exceeds that on the B–C links by about

$$\frac{5·5}{103} \times 100 = 5·3\%$$

Taking account of calls abandoned before dialling, and false traffic due to receivers being accidently left off the hook, the difference may be somewhat greater. Delay in clearing blocked calls is another factor which contributes to traffic differences between stages.

Delays before dialling have a similar effect in the case of link systems. Since the main switching path is set up in one logical operation, however, any differences in holding time between successive ranks will be mainly due to the operating characteristics of the system rather than the subscribers' behaviour, and they are likely to be small, probably negligible in terms of traffic flow.

Allowing for these differences, the traffic flow throughout a network must of course be consistent. If we regard the *effective* holding time of a call as beginning when it is connected to the called line, then the total effective traffic entering and leaving any switching point during a given period, must be equal. This does not always apply in traffic calculation, because it may be necessary to allow for differences in group busy hours, so that the traffic flows used for equipment provisioning may not balance at every point.

It is common practice to characterise the traffic demand on a network by the average traffic in erlangs per subscriber's line, including both originated and received calls. Care should be taken in relating this figure to the traffic carried by each switching stage, to ensure that the same calls are not counted twice. Confusion is unlikely to arise when considering the traffic in the whole network, but may do so when, as is normally the case in practical traffic engineering, part of the network has to be considered in isolation. As an example, Fig. 18 shows a 1000-line exchange, in which each line originates 0·05 E on the average, and receives 0·04 E. For simplicity, these figures will be taken as referring to effective traffic, in the sense explained above. It should be noted, however, that this is not the usual interpretation; the quoted originated traffic generally includes dialling delays etc. The difference between originated and received effective traffic per line implies a difference between traffic from and to other exchanges. Let us suppose that 0·02 E, on the average, terminates at the same exchange, so that 0·03 E goes to a distant exchange and 0·02 E comes from a distant exchange. Then, on the average, there are 30 E of outward traffic, 20 E inward, and 20 E terminating at the same exchange, a total of 70 E. If, how-

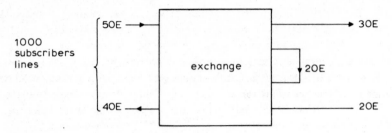

Fig. 18 Traffic circulation in exchange

ever, we consider the subscribers' lines rather than the network, the total traffic is

$$1000(0.04 + 0.05) = 90\,E$$

of which 20 E represents 'own exchange' calls which are counted twice.

If the subscribers' lines are connected to different sets of links for originated and received calls, as in Figs. 2 and 3, these links carry 50 E and 40 E, respectively. If there is a common group of links, as in Fig. 6, it is a matter of convenience whether they are said to carry 90 E or 70 E, the important point is to remember that part of this traffic consists of calls occupying two links, and to make an allowance for this in traffic calculation (Section 3.6).

2.7 Random bursts

It is sometimes useful to calculate the frequency with which an exceptionally high incidence of calls is likely to occur as a result of random variation in a Poisson process. Let us define a burst as the arrival of more than a specified number of calls (x) during a specified period (T). For convenience, the average holding time is taken as unity, so that the average number of arrivals per unit time is the traffic flow in erlangs A.

Consider a particular call arrival. The probability that x calls have arrived within the period T immediately preceding this arrival, thus producing a 'burst' along with the call which ends the period is

$$e^{-AT}\frac{(AT)^x}{x!}$$

During a very long period Z, the average number of call arrivals is AZ, of which

$$AZe^{-AT}\frac{(AT)^x}{x!}$$

produce bursts. The average interval between successive bursts is obtained by dividing this expression into Z giving

$$\left[Ae^{-AT} \frac{(AT)^x}{x!} \right]^{-1}$$

The following method, which is sometimes used, is incorrect. Let us divide Z into Z/T intervals of duration T. On the average, the number of bursts, or intervals containing more than x call arrivals is

$$Z/T \sum_{r=x+1}^{\infty} \frac{(AT)^r}{r!} e^{-AT}$$

Hence the mean time between bursts is

$$\left(\frac{\text{total number of intervals}}{\text{number of bursts}} \right) \times T$$

$$= \left(\sum_{r=x+1}^{\infty} \frac{(AT)^r}{r!} e^{-AT} \right)^{-1} \times T \text{ average holding times}$$

For example let $A = 1, T = 10, x = 19$

The first formula gives a mean time between bursts of 268 holding times, while the second one gives 2895 holding times. In fact, the second method gives the correct answer to a different problem, which may be stated as follows. A recording device counts the number of calls in successive, nonoverlapping intervals T. What is the mean time between successive counts of more than x calls?

2.8 Grade of service

If just sufficient equipment were provided to carry the average traffic flow, an unacceptably high proportion of calls would encounter congestion. On the other hand, increasing the provision of equipment beyond a certain limit does not produce any significant improvement in service. To establish a standard basis of equipment provision, it is necessary to devise a measure of service to quantify the inconvenience suffered by subscribers as a result of congestion. The definition of inconvenience depends on how the system deals with calls which encounter congestion. In a loss system, busy tone is immediately returned to the caller, who must replace his receiver before trying again. In a delay system, the caller can wait, possibly for a limited time, during which he is immediately connected when a path is available; he is only considered to be inconvenienced if the delay exceeds some maximum tolerable value. Telephone systems generally employ a mixture of loss and delay conditions. Congestion at the first switching stage (e.g. the links connecting the A and B switches in Fig. 2) merely delays the return of dialling tone, whereas busy tone is given in the event of congestion at a later stage. To provide delay facilities throughout would tend to aggravate congestion under heavy traffic conditions by overloading the system with waiting calls. This does not apply so much to message switching systems of the store and forward type used in data communication, in which each link is released as the message is passed on to the next stage, and may enter a queue at each step.

In step-by-step systems, it is usual to specify the allowable probability of loss at each stage; this, of course, is equivalent to the proportion of calls which are lost in the long run. The specified loss for each trunk group refers to the busy hour for that group. For convenience, delays before dialling are often treated as losses. This can be regarded as a measure of service as seen from the point of view of an average subscriber. The service actually experienced by any particular subscriber, however, depends on a number of different factors. For example, one who makes few calls and distributes them evenly during the day will suffer less inconvenience than one who makes heavy use of the telephone during the busy hour. It is hardly practicable however, to take account of such diverse circumstances in grade-of-service calculations.

The specified loss per stage must be low enough to ensure that a long-distance call switched through a number of exchanges has a reasonable grade of service. The allowable loss is usually in the range 0·001 to 0·01 per stage for internal links, and 0·005 to 0·05 for expensive external links. The lower figure in each range usually requires between 10% and 40% more links than the higher one.

These figures apply to average busy-hour traffic. Day-to-day variations, however, have the effect of increasing the average loss rate. For example, a 10% increase in traffic might increase the probability of loss from, say, 0·001 to 0·005, whereas a 10% reduction obviously cannot reduce it by more than 0·001. To guard against excessive degradation of service under overloading, it is usual to specify the maximum loss for a given increase in traffic. Typical figures are twice or five times the normal loss rate when the traffic increases by 10% or 20%, respectively. If necessary, the normal design loading is reduced to meet these criteria.

At low blocking, the total probability of loss is approximately the sum of the probabilities for each switching stage (Section 5.1). For example, an own-exchange call might traverse three switching stages at loss 0·005 each, giving a total loss of 0·015, while a call to a distant exchange, via one transit exchange, might traverse four internal stages at 0·005 and two junction routes at 0·02, a total of 0·06. In the case of conditional-selection systems, losses cannot be precisely allocated to each stage, since they result from a combination of states at different stages; it is therefore necessary to specify the total loss for any sequence of stages forming a link system with conditional selection.

These figures should not be interpreted too literally, as there are a number of factors which tend to reduce the total congestion well below its nominal value. The traffic on most routes is growing continuously, whereas extension of equipment takes place in steps, each of which should be designed to allow for future growth. Thus, the actual grade of service should be better than the nominal one most of the time, although degradation may occur when a trunk group reaches the limit of its extensible capacity, or when there is a delay in the supply of equipment. Again, different trunk groups may require extension at different times, and may have different busy hours, which tends to reduce end-to-end congestion. Overloads are unlikely to occur on all groups simultaneously.

The figures quoted do not include call failures due to faulty equipment. Tolerable fault rates are usually specified in terms of mean time between failures, in accordance with normal reliability-engineering practice, rather than probability of blocking. One reason for this is that, while the latter is a good measure of service from the subscriber's point of view, it can be misleading in other respects. For example, if one call in 1000 were to blow a fuse, very little inconvenience might be caused to subscribers, but a great deal to the maintenance staff of a large exchange, who might have to replace fuses every few minutes. In the case of faults which reduce traffic capacity without complete loss of service to any route, however, the tolerable probability of blocking in the event of a fault may also be specified. This may be related to the mean time between failures, higher blocking being tolerated for less frequent faults.

Standard grades of service have usually been arrived at on the basis of judgement and experience, the aim being to achieve the best service obtainable with the technology available at reasonable cost. The standards should not be too inflexible, and may need to be reviewed from time to time in the light of technical developments. Even if a new system is required to meet the same overall standard of service as the existing one, detailed application of present grade-of-service specifications may not necessarily be the best way of achieving this end.

In principle, the optimum provision of equipment is achieved when the benefit of any further improvement in service, including increased revenue and any indirect social benefits which are relevant to the policy of the administration, is just balanced by the cost of providing it; cost and benefits being expressed in the same units. This concept of 'marginal utility' was applied to telephone traffic by K. Moe as early as 1923;[1] an example is given in Section 3.9.

An alternative approach to the problem of service optimisation was suggested by de Ferra and Massetti;[67] this maximises the overall standard of service subject only to the willingness of the community to pay for improvement. The procedure can be summarised as follows. The overall grade of service is expressed in terms of a mathematical function Q, which is designed to take account of total losses and subscribers' reaction to different loss rates. Q ranges from 0 to 1 as service improves. Denoting the cost of the network by C, an increase in C produces an increase in Q, but the rate of improvement (dQ/dC) eventually falls off. There is a certain rate at which the community is no longer willing to pay for further improvement. This is given by

$$dQ/dC = K$$

The value of K is not known accurately, but it can be estimated approximately as follows. C and Q are calculated for the present network. The network is then re-costed using, say, twice the present loss probabilities for every stage, but preserving the same ratios of loss throughout. The process is repeated with other factors or multiples of the present losses, and a Q/C curve is drawn. The slope of the curve at the point representing the present network is taken as the value of K. In other words, it is assumed that, while the present network may not be optimal, the fact that it is designed to a certain grade of service rather than half or twice that grade, reflects the value which the community attaches to the service as such, not merely to the present system. Using this value of K, the maximum Q network can be calculated, which may, in fact, have a lower cost than the present network. For example, in a typical case, optimisation reduced the mean loss from 0·1344 to 0·1240, and the total cost by 0·9%. Alternatively, using different values of K, the mean loss could be reduced to 0·1061 with no change in cost, or the cost by 1·38% with no change in grade of service.

In the past, the difficulty of establishing reliable economic parameters accurately, and computational difficulties, have discouraged the general adoption of econometric methods of service optimisation, in spite of Moe's pioneer work. Today, however, there is a revival of interest in such methods. Modern computers make the calculations feasible, while the speed of technological change requires new approaches if the best use of new techniques is to be ensured.

Chapter 3

Lost-call theory for full-availability groups

The *availability* of the outlets of a switching network is the number of outlets, in the required route, which can be reached from an inlet. Full availability exists when any free inlet can reach any free outlet, regardless of the state of the system. Thus, in Fig. 2, the ten links between switches A_1 and B form a full-availability group with respect to the inlets of A_1, and the ten links between B and C_1 form a full-availability group with respect to the inlets of B.

It is easily seen that the efficiency of a full-availability group increases with its size. For example, consider two groups of ten links each, the average traffic being the same in both cases. One group can be congested while the other has free links. If, however, the two are combined into a single full-availability group of 20 links, serving all inlets, there can be no congestion until all the links are busy.

Since the large group has a lower probability of congestion than the two small ones for a given traffic flow, it can carry more traffic at a given probability of congestion. On the other hand, this renders the large group more sensitive to overloading, since there is less margin of idle time to absorb an increase in traffic.

These principles are illustrated in the table, which shows the traffic capacity of 160 links arranged as one or more full-availability groups: the stipulated probability of loss is 0·005 for normal traffic, with the proviso that the loss after 10% overload shall not exceed 0·01. The bracketed figures in the third column show the percentage increase in normal traffic capacity when the availability is doubled. Thus, an increase in availability from 10 to 20 produces an increase in traffic capacity from 63·4 E to 88·7 E, i.e. 40%. As the availability increases the rate of increase in capacity falls off, so that, at some point depending on equipment costs, the benefit of increased availability is outweighed by the cost of the larger switches necessary to provide it. At availability 10, the overload grade of service corresponds to a higher traffic (before a 10% increase) than the normal grade of service; hence, the latter can be used as the basis of design, since the overload proviso is automatically met. At higher availabilities, however, the system must be underloaded at normal traffic, giving a loss of less than 0·005, to ensure that a 10% increase does not raise the loss

Table 1   Traffic capacity of various full-availability groupings of 160 links

Number of groups	Size of group	Total traffic in erlangs Loss 0·005	Loss 0·01	Loss 0·01 after 10% overload	Design-traffic load
16	10	63·4	71·4	64·9	63·4
8	20	88·7 (40%)	96·2	87·5	87·5
4	40	109·5 (23%)	116·0	105·5	105·5
2	80	125·3 (14%)	130·7	118·8	118·8
1	160	136·8 ((9%)	141·2	128·4	128·4

above 0·01. The design traffic is shown in the sixth column, which is the lower of the third and fifth columns. The table is based on Erlang's lost-call formula, which is dealt with in the next Section.

3.1 Poisson input with lost calls cleared

The full-availability group is the simplest traffic system of more than one trunk, and was naturally the first to be investigated. One of the earliest attempts to determine the probability of blocking was made by W.H. Grinsted in 1907, although not published until 1915.[2] Assuming Poisson distribution for call arrivals, and equal holding times for all calls, he argued that blocking occurs if, and only if, the number of calls offered, during a preceding period of one holding time is greater than the number of trunks. The probability of this event is

$$\sum_{x = N + 1}^{\infty} e^{-A} \frac{A^x}{x!}$$

where A = average number of calls offered per holding time (i.e. traffic offered in erlangs), and N = number of trunks.

    This formula does not, however, give the probability of congestion in the sense required for calculating the grade of service, i.e. the probability of finding the group congested at any given time, when a call is offered. This is illustrated in Fig. 19, which shows a possible sequence of events in a group of two trunks during a period of 13 min, the call-holding time being 4 min. A call made at the end of the eighth minute finds a trunk free, despite the fact that more than two calls have arrived during the preceding 4 min. Two of these, however, have been blocked by previous calls, with·the result that the group is no longer congested at the end of this period. A rigorous solution of the problem had to await the formulation of the concept of statistical equilibrium by A.K. Erlang, who thereby laid the foundation of the modern theory of stochastic processes.[3]

Erlang made the following assumptions:

(*a*) Calls occur individually and collectively at random, i.e. in accordance with the Poisson distribution. This implies a very large, theoretically infinite number of calling sources, as we have seen.

(*b*) A state of statistical equilibrium exists.

(*c*) Calls originating when all trunks are busy are lost, and their holding time is zero. This assumption, which is usually referred to as 'lost calls cleared', ignores the effect of repeated attempts by unsuccessful callers, but this is not a serious source of error at normal grades of service.

(*d*) To simplify the following proof, negative-exponential holding-time distribution is assumed. The theory holds good, however, for any holding-time distribution, as Erlang claimed, although without satisfactory proof; this was later given by Kosten.[68]

The direction of the traffic and the order in which trunks are chosen are immaterial. If, however, the group carries traffic in both directions, every trunk must be capable of use in either direction as required.

Let us consider a very short interval dt, at the beginning of which x trunks are busy. The interval may contain a call arrival or termination, but the chance of more than one event, being proportional to dt^2 or higher powers, is negligible. It is convenient to take one average holding time as the unit of time. By assumption (*d*), each call has probability e^{-dt} of outlasting dt, irrespective of its previous duration. The probability that all calls will outlast dt is $(e^{-dt})^x$ and the probability of a call termination in dt is therefore $1 - e^{-xdt} = 1 - (1 - xdt \pm$ terms in higher powers of $dt) \doteqdot xdt$. By assumption (*a*), the probability of a call arrival is Adt (Section 2.4).

Let $p(x)$ denote the probability of finding x trunks busy simultaneously. By assumption (*b*), $p(x)$ is independent of the time at which the system is examined, provided there is no knowledge of its past states which conveys information about its probable present state. This implies, roughly speaking, that the state 'x trunks busy' is created as often as it is destroyed during a long period; otherwise, the probability of finding it would change with time.

Let $[x]$ denote the state 'x trunks busy'. The system can enter the state $[0]$

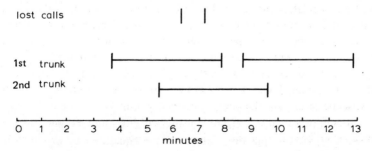

Fig. 19 Traffic in a full-availability group

only as a result of a call termination, and can leave it only as the result of a call arrival. It follows that [0] changes to [1] as often as [1] changes to [0]. Similarly, [1] changes to [2] as often as [2] changes to [1]; otherwise, destructions of state [1], i.e. changes from [1] to [0] and [1] to [2], would not balance creations, i.e. changes from [0] to [1] and [2] to [1]. Hence, in general, if $0 \leqslant x < N$, [x] changes to [x + 1] as often as [x + 1] changes to [x], so the probability of finding either of these events in a short interval dt is the same.

The probability that the interval contains a transition from [x] to [x + 1], i.e. that the system is in state [x] at the beginning of the interval and a call arrives during it, is $p(x)A dt$. Similarly, the probability that it contains a transition from [x + 1] to [x] is $p(x + 1)(x + 1)dt$. Hence

$$p(x + 1) = \frac{A}{(x + 1)} p(x)$$

giving the following set of equations:

$$p(1) = Ap(0)$$

$$p(2) = \frac{A}{2} p(1) = \frac{A^2}{2!} p(0)$$

$$p(3) = \frac{A}{3} p(2) = \frac{A^3}{3!} p(0)$$

In general, $$p(x) = \frac{A^x}{x!} p(0) \qquad (0 \leqslant x \leqslant N)$$

Moreover, $$p(0) + p(1) + p(2) + ... + p(N) = 1$$

$$p(0) = 1 \left/ \left(1 + A + \frac{A^2}{2!} + ... + \frac{A^N}{N!} \right) \right.$$

$$p(x) = \frac{A^x}{x!} \left/ \left(1 + A + \frac{A^2}{2!} + ... + \frac{A^N}{N!} \right) \right.$$

This is the Erlang distribution.

Since the arrival of calls is unrelated to the state of the system [assumption (*a*)], the *call congestion*, i.e. the proportion of lost calls in the long run, is equal to the probability of finding all trunks busy, which is usually denoted

$$E_{1,n}(A) = \frac{A^N}{N!} \left/ \left(1 + \frac{A^1}{1!} + \frac{A^2}{2!} + ... + \frac{A^N}{N!} \right) \right.$$

This is Erláng's lost-call formula. The same expression gives the *time congestion*, i.e. the proportion of time during which all trunks are busy during the busy hour. Thus, with Poisson input, call and time congestion are equal, but as will be seen later, this is not true in general.

As N increases, the denominator approximates closely to e^A, so that the number of simultaneous calls tends to have a Poisson distribution when congestion is low.

3.2 Alternative derivation of state equations

A more general approach to the formulation of equations of state, which leads to the same results, is as follows:

Let $p(x, t)$ = the probability that x trunks are busy at time t.

The state 'x trunks busy at $t + dt$' must be preceded, at time t and during dt, by one of the following combinations of states and events, neglecting the possibility of more than one event in dt.

$x - 1$ calls exist at t, one call arrives in $dt(x > 0)$; probability $= p(x - 1, t)Adt$.

x calls exist at t, no change in dt; probability $= p(x, t)(1 - Adt - xdt)$.

$x + 1$ calls exist at t, one terminates in $dt(x < N)$; probability $= p(x + 1, t)(x + 1)dt$.

Equating the probability of the state 'x calls at $t + dt$' to the total probabilities of the preceding states and events from which it can arise, we have

$$p(x, t + dt) = p(x - 1, t)Adt + p(x, t)(1 - Adt - xdt) + p(x + 1, t)(x + 1)dt$$

Thus,

$$p(x, t + dt) - p(x, t) = p(x - 1, t)Adt - p(x, t)(A + x)dt + p(x + 1, t)(x + 1)dt$$

Dividing by dt and letting $dt \to 0$, this becomes

$$\frac{\partial p(x, t)}{\partial t} = p(x - 1, t)A - p(x, t)(A + x) + p(x + 1, t)(x + 1)$$

By assumption (b), however, $\partial p(x, t)/\partial t = 0$, and the p-terms are independent of t. Hence

$$p(x)(A + x) = p(x - 1)A + p(x + 1)(x + 1)$$

$$\text{with } 0 \leqslant x < N \quad \text{and} \quad p(-1) = p(N + 1) = 0$$

The left-hand side of the equation represents the rate at which the system leaves the state $[x]$, and the right represents the rate at which it enters this state. As before we have

$$Ap(0) = p(1)(1)$$

Thus
$$p(1) = Ap(0)$$

Similarly
$$2p(2) = (A + 1)p(1) - Ap(0) = A^2 p(0)$$

Thus
$$p(2) = \frac{A^2}{2!} p(0) \quad \text{etc.}$$

This method, unlike that of Section 3.1, does not require that the rate of transitions between each pair of states shall be equal in both directions; as will be seen later, this is not always the case. Moreover, the method can be generalised to non-equilibrium conditions.

3.3 Arrangements of calls in a trunk group

The concept of statistical equilibrium applies not only to states defined by the total number of calls in progress, but also to particular arrangements of calls on the trunks, or sets of arrangements. For example, let us arbitrarily define the state A of a group of ten trunks as the state in which there are three calls in progress, occupying either trunk nos. 1, 3 and 10, or 5, 6 and 7. If the system is in equilibrium, the probability of finding it in state A is independent of the time at which it is examined.

Let B denote some other state which can change to A as the result of an arrival or termination; then it may or not be the case that the rate of transition form A to B is equal to the rate of transition from B to A. As a simpler example, consider a 2-trunk full-availability group of trunks, which are always hunted in the same order, and let [10] denote the state with choice no. 1 busy and choice no. 2 free etc. Clearly there can be no transitions from [00] to [01], although transition from [01] to [00] can occur through a call termination.

As in Section 3.2 the state probabilities can be determined by equating the rates of transition into and out of each state. If $p(10)$ denotes the probability of state [10], we have

$$Ap(00) \; = \; 1p(10) + 1p(01)$$

$$(A + 1)p(10) \; = \; Ap(00) + 1p(11)$$

$$(A + 1)p(01) \; = \; 1p(11)$$

$$2p(11) \; = \; Ap(01) + Ap(10)$$

These equations, together with the condition that the sum of the probabilities of all states is unity, are sufficient to determine the probabilities.

If the trunks are tested in random order, the assumption of equal transition rates in both directions between each pair of states may be justified, at least as a practical approximation. Strictly speaking, however, this may require more thorough randomisation than is usually practicable. A common method of selection is for a switch to test first the trunk at which it happens to be standing, and, if that is busy, to hunt sequentially from that position until a free trunk is found. This tends to produce a gradual cyclic drift of busy trunks in the direction of hunting, so that the rate of entering or leaving a given state may change cyclically with time.

3.4 Sequential hunting

While the order in which trunks are hunted in a full-availability group does not affect the probability of congestion, it may affect their relative loading. If they are hunted sequentially from a fixed position, as with Strowger selectors, the state of the first k consecutive choices ($1 \leqslant k < N$) is obviously unaffected by that of any subsequent choices; so that the early choices can be treated as a self-contained full-availability group of k trunks. The traffic offered to the $(k + 1)$th choice, i.e. the overflow from the first k choices, is $AE_{1,k}(A)$. The traffic carried by the kth choice is the difference between the traffic offered to the kth and $(k + 1)$th choices, i.e.

$$a_k = A \{E_{1,k-1}(A) - E_{1,k}(A)\}$$

This may be expressed in the form

$$a_k = A \left\{ \frac{E_{1,k}(A)(k/A) \sum\limits_{r=0}^{k} \dfrac{A^r}{r!}}{\sum\limits_{r=0}^{k} \dfrac{A^r}{r!} - \dfrac{A^k}{k!}} - E_{1,k}(A) \right\}$$

Dividing the numerator and the denominator of the first term in brackets by

$$\sum_{r=0}^{k} \frac{A^r}{r!}$$

we have

$$a_N = E_{1,N}(A)(N-A)$$

If the probability of loss $E_{1,N}(A)$ is small, the traffic carried by the last trunk ($k = N$) is approximately

$$a_k = E_{1,n}(A)(N-A)$$

The traffic carried by the first choice is

$$a_1 = \frac{A}{1+A} \left\{ \frac{1}{1 - \dfrac{A}{1+A}} - A \right\} = \frac{A}{1+A} = E_{1,1}(A)$$

This result can be proved directly from assumption (a), since a_1 and $E_{1,1}(A)$ are, by definition, the time and call congestion, respectively, for the first trunk, which are equal for Poisson input. In the case of later choices, however, they are not equal. The probability b_k that choice no. k is busy when a call is offered to it is the ratio of the traffic rejected by that choice to the traffic offered to it, i.e.

$$b_k = \frac{E_{1,k}(A)}{E_{1,k-1}(A)}$$

It can be shown by rearrangement of terms that

$$a_k = AE_{1,k-1}(A)(1 - b_k)$$

In other words, the traffic carried by the kth trunk is the product of the traffic offered to it and the conditional probability that the choice is free, given that a call is offered, as would be expected. It can be proved that $a_k < b_k$ if $k > 1$. This would be expected intuitively, since the arrival of calls at the kth choice is concentrated into periods when the first $k - 1$ choices are all busy. This produces a pattern of sharp peaks, so that, after each seizure of the trunk, there is a higher than average chance of further attempts to occupy it.

As an example, if 4·0 E is offered to 10 trunks with sequential hunting, the probability of blocking is 0·0053. The first and last choices carry 0·8 E and 0·0318 E, respectively. The probabilities of these choices being busy when a call is offered are 0·80 and 0·40, respectively. Thus, although the last choice is only in use for 3·18% of the busy hour, less than 2 min, 40% of the calls offered to it are rejected. The absence of the tenth choice would more than double the probability of congestion to 0·013.

3.5 Limited sources with lost calls cleared

When the number is relatively small, the proportion of free sources, and therefore the probability of call arrival, may fluctuate considerably with the state of the traffic, so the Poisson distribution is not applicable. The theoretical solution of this case is chiefly attributed to Engset[26] and Martin.[69] The following assumptions are made:

(a) A free source is equally likely to originate a call at any time, irrespective of the state of the system.

(b), (c) and (d) are the same as for Erlang's formula.

(e) All sources originate the same average traffic.

(f) All sources can make but not receive calls; this is realistic in certain circumstances, e.g. trunk groups serving coin-box lines exclusively. In practice, however, the theory is applicable as an approximation in other cases, as will be shown later.

Let N = number of trunks

  M = number of sources

  A = traffic offered, in erlangs, by all sources

  a = calling rate per free source; i.e. the probability that a source which is free at the beginning of a short interval dt makes a call during that interval is $a\,dt$

  B = probability that a call is lost (call congestion)

Let $p(x)$ = probability of finding x busy trunks at a random instant.

If x trunks are busy, the probability of a call arrival during dt is $(M-x)\,adt$, while that of a call termination is xdt. The equations of statistical equilibrium are

$$p(x+1)(x+1)\,dt = p(x)(M-x)\,adt \qquad (0 \leqslant x < N)$$

Hence
$$p(1) = Map(0)$$

$$p(2) = \frac{(M-1)a}{2}p(1) = \frac{(M-1)Ma^2}{2}p(0)$$

$$p(3) = \frac{(M-2)(M-1)Ma^3}{3!}p(0)$$

In general,
$$p(x) = \frac{(M-x+1)(M-x+2)\dots(M-1)Ma^x}{x!}p(0)$$

$$\equiv \binom{M}{x}a^x p(0)$$

$$\equiv \frac{\binom{M}{x}a^x}{\left\{1 + \binom{M}{1}a^1 + \binom{M}{2}a^2 + \dots + \binom{M}{N}a^N\right\}}$$

since
$$p(0) + p(1) + \dots + p(N) = 1$$

This is the Engset distribution. The time congestion is $p(N)$. The call congestion is derived as follows: The probability that N trunks are busy at the beginning of a random interval dt, and that a call arrives during dt, is

$$p(N)(M-N)\,adt$$

This probability can also be expressed as the product of the unconditional probability that a call occurs during dt, and the conditional probability that, if a call occurs, it is lost; the second term is the call congestion, by definition. The first term is Adt; this follows from assumption (a), in the same way as for Poisson input (Section 2.4). Hence

$$\frac{\binom{M}{N}a^N(M-N)\,adt}{\displaystyle\sum_{r=0}^{N}\binom{M}{r}a^r} = AdtB$$

Thus
$$B = \frac{M\binom{M-1}{N}a^{N+1}}{A\displaystyle\sum_{r=0}^{N}\binom{M}{r}a^r}$$

A more convenient form for computation can be obtained as follows: The event 'a call arrives during dt' comprises the following mutually exclusive events:

0 calls exist at the beginning of dt; a call arrives during dt:
1 exists at the beginning of dt; a call arrives during dt;

and so on.

Equating probabilities we have

$$Adt = p(0)Madt + p(1)(M-1)adt + ... + p(r)(M-r)adt + ... + p(N)(M-N)adt$$

$$A = \frac{\sum_{r=0}^{N} \binom{M}{r} a^r (M-r)adt}{\sum_{r=0}^{N} \binom{M}{r} a^r} = \frac{M \sum_{r=0}^{N} \binom{M-1}{r} a^{r+1}}{\sum_{r=0}^{N} \binom{M}{r} a^r}$$

Thus

$$B = \frac{\binom{M-1}{N} a^N}{\sum_{r=0}^{N} \binom{M-1}{r} a^r}$$

In practice, a is an inconvenient quantity to determine; it can be eliminated as follows. Each source offers A/M erlangs, of which $(A/M)(1-B)$ erlangs is carried. The probability that a source is free at a random instant is therefore

$$1 - \frac{A}{M}(1-B)$$

If nothing is known about the state of a source, the probability that it offers a call during dt is

$$\frac{A}{M}dt$$

This can be regarded as the product of the probability that the source is free and conditional probability that, if free, it offers a call during dt.

Therefore

$$\frac{A}{M}dt = \left\{ 1 - \frac{A(1-B)}{M} \right\} adt$$

Thus

$$a = \frac{A}{M - A(1-B)}$$

Finally, if B_N denotes the call congestion for N trunks, we have

$$B_N = \frac{(M-N)aB_{N-1}}{N + (M-N)aB_{N-1}}$$

Thus, having calculated B_1 for given values of a and M, other B_N terms can be obtained by iteration, and A can be calculated in terms of a, M and B_N; thence tables of B_N against A for given values of M can be constructed, a being eliminated.

If M is very large, a is approximately A/M and $\binom{M}{r}$ is approximately $\dfrac{M^r}{r!}$ so that the Engset distribution approximates to the Erlang.

Since the probability of a call arrival with a limited number of sources decreases as the number of busy trunks increases, blocking is less than that with Poisson input for a given traffic and number of trunks. As an example, the call congestion values when a traffic of 12·0 E is offered to 20 trunks by various numbers of sources are as follows

Table 2

Number of sources	Call congestion
20	0
30	0·0011
40	0·0027
50	0·0039
100	0·0068
200	0·0083
∞	0·0098

If $N = M$, clearly $B = 0$ and

$$p(x) = \frac{\binom{M}{x}\left(\dfrac{A}{M-A}\right)^x}{\left(1 + \dfrac{A}{M-A}\right)^N} = \binom{M}{x}\left(\frac{A}{M}\right)^x \left(1 - \frac{A}{M}\right)^{M-x}$$

so the Engset distribution reduces to the binomial. The variance of the traffic offered by a limited number of sources is therefore

$$M\left(\frac{A}{M}\right)\left(1 - \frac{A}{M}\right) = A\left(1 - \frac{A}{M}\right)$$

which is less than A, the variance of the same volume of Poisson traffic. For this reason, limited source traffic is sometimes described as 'smooth', the term 'pure chance' being reserved for Poisson traffic. Some authorities, however, refer to the latter as 'pure chance traffic no. 1' and the former as 'pure chance traffic no. 2', (p.c.t. 1, p.c.t. 2), the term 'smooth traffic' being reserved for traffic smoothed by the switching arrangements (Section 5.2).

Uneven loading of sources tends to reduce the variance further. As a simple example, suppose there are two classes of sources, one comprising m sources originating a total of y erlangs, the other comprising $M - m$ sources and originating a total of $A - y$ erlangs. The variance of the combined traffic is the sum of the separate variances, which is

$$
y \left(1 - \frac{y}{m} \right) + (A - y) \left(1 - \frac{A - y}{M - m} \right)
$$

$$
= A - \left\{ \frac{y^2}{m} + \frac{(A - y)^2}{M - m} \right\}
$$

$$
= A - \frac{A^2}{M} \left\{ \frac{\left(\dfrac{My}{A} - m \right)^2}{m(M - m)} + 1 \right\} \leqslant A \left(1 - \frac{A}{M} \right)
$$

Equality obtains only if $y/m = A/M$ in which case all sources have the same calling rate.

Fig. 20 shows an example of a system where the Engset distribution may be applicable as an approximation, although the assumptions, (f) in particular, are not all satisfied. M sources are connected to a switching network, such as a p.a.b.x.; internal calls, i.e. calls between any of the M sources, are carried on a group of N_I trunks, outgoing calls on a group of N_O trunks, and incoming calls on a group of N_C trunks. The latter are assumed to come from an effectively infinite number of sources, so that they have a Poisson distribution. Full availability exists in all cases. In the case of incoming calls, it is assumed that calls to busy sources (or, in this context, more correctly 'sinks') have zero holding time, so that the traffic distribution is effectively determined by successful calls, and the limited number of sinks has a similar smoothing effect to that of the limited number of sources on outgoing calls. In applying the limited source congestion formula, the number of sources is taken as M for both incoming and outgoing traffic, and as $M/2$ for internal traffic. The formula tends to over-estimate congestion, since it assumes that any source which is not using the route under consideration is free to make a call, or receive one, whereas it may be already busy on another route. More accurate formulas for this situation have been developed by Rubas[70] and Herzog.[71]

3.6 Bothway trunks with internal traffic

Fig. 21 illustrates a case where two trunks in a bothway group may be required simultaneously by the same call. Instead of connecting all subscribers' lines direct to the exchange, a number of subscribers in a neighbourhood may be connected to a 'line concentrator', which gives access to the exchange via a relatively small number of trunks, thus saving the cost of a separate exchange line per subscriber. The concentrator has no switching capability and internal calls (between subscribers

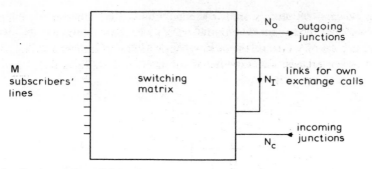

Fig. 20 Application of Engset's formula

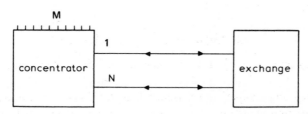

Fig. 21 Line concentrator

connected to the same concentrator) are switched through the exchange, thus requiring two trunks to connect the calling and called subscribers to the exchange. The trunks are fully available and can be used in either direction as required. Blocking in the exchange or other parts of the network will be neglected. The holding time of a trunk in the outward direction includes dialling time, whereas an inward trunk cannot be seized until the destination code has been dialled.

Let c_I, c_E = average number of internal and external calls offered per unit time

N = number of trunks connecting the line concentrator to the exchange for switching

M = number of sources (subscribers' lines) connected to the line concentrator

h_I, h_E = trunk holding times averaged over both directions for internal and external calls, respectively

$p(x, y)$ = probability that x trunks are held by internal and y by external calls.

For simplicity, it is assumed that the two trunks used by an internal call are seized at the same moment, so that x must always be even; this is a reasonable approximation since the difference between the holding times of the inward and outward trunks is usually small compared with their average value. It is also assumed that M is large enough to justify regarding the input as Poissonian; for the general case see Bazlen.[72]

As usual, equations of statistical equilibrium are derived by equating the probabilities of the system entering and leaving each state during a short interval dt. The probability that one of the $x/2$ internal calls terminates, releasing two trunks, is $xdt/2h_I$. The equations are as follows, omitting dt:

$$\left(c_I + c_E + \frac{x}{2h_I} + \frac{y}{h_E}\right)p(x,y) = c_I p(x-2,y) + c_E p(x,y-1)$$

$$+ \frac{x+2}{2h_I}p(x+2,y) + \frac{y+1}{h_E}p(x,y+1)$$

with
$$0 \leqslant x \leqslant N, x = 2n \quad (n \text{ an integer})$$

$$0 \leqslant y \leqslant N, p(x,y) = 0 \quad \text{if } x+y > N$$

$$\sum_{x=0}^{N} \sum_{y=0}^{N} p(x,y) = 1$$

If $x+y=N$, the terms c_I and c_E are omitted on the left-hand side. These equations are sufficient to determine all state probabilities. A more convenient relationship can, however, be obtained by assuming equilibrium in a narrower sense, i.e. equal probability of transition in both directions between pairs of states[73] (Section 3.3). This leads to the following equations:

$$c_E p(0,y) = \frac{y+1}{h_E}p(0,y+1)$$

$$c_E p(2,y) = \frac{y+1}{h_E}p(2,y+1)$$

etc.

whence
$$p(0,1) = \frac{h_E c_E p(0,0)}{1}$$

$$p(0,2) = \frac{h_E c_E p(0,1)}{2}$$

$$= \frac{h_E^2 c_E^2 p(0,0)}{2!}$$

or, in general,
$$p(0,y) = \frac{h_E^y c_E^y p(0,0)}{y!} \quad (0 \leqslant y \leqslant N)$$

Similarly
$$p(x,y) = \frac{h_E^y c_E^y p(x,0)}{y!} \quad (0 \leqslant y \leqslant N-x)$$

A similar set of equations is obtained from transitions due to arrival and termination of internal calls, giving

$$c_I p(x, 0) = \frac{x + 2}{2h_I} p(x + 2, 0)$$

from which we obtain $\qquad p(x, 0) = \frac{h_I^{x/2} c_I^{x/2} p(0, 0)}{(x/2)!} \qquad (0 \leqslant x \leqslant N)$

Thus $\qquad p(x, y) = \frac{(h_I c_I)^{x/2} (h_E c_E)^y}{(x/2)! \, y!} p(0, 0) \qquad (0 \leqslant x + y \leqslant N, x \text{ even})$

Thus the $p(x, y)$ terms can be calculated in terms of $p(0, 0)$, the value of which is determined by the condition that

$$\sum_{x=0}^{N} \sum_{y=0}^{N} p(x, y) = 1$$

The probability of congestion for an internal call is

$$\sum_{x=0}^{N} p(x, N - x) + \sum_{x=0}^{N-1} p(x, N - x - 1)$$

3.7 Poisson input with lost calls held

Molina's 'lost-calls-held' formula is based on the same assumptions as Erlang's 'lost-calls-cleared' formula, except that assumption (c) is replaced by the following admittedly artificial assumption, which was introduced for mathematical simplicity.[27]

When a call finds all trunks busy it continues to demand service for a period equal to the holding time it would have had if successful. If, during this period, a trunk becomes free, it is seized and rendered unavailable for the unelapsed part of the holding time, but the call is still regarded as lost.

Let x denote, not the number of busy trunks, but the total number of calls in the system, including waiting calls and calls receiving truncated service after waiting. Clearly, x can take any value from 0 to infinity in theory. As the total time for which a call is in the system is the same whether or not it is successful, the probability of a call termination during dt is $x \, dt$; 'termination' means either the release of a trunk, or the abandonment of a waiting call which has not yet found a trunk. The probability of a call arrival is $A \, dt$ as before. Hence, the equations of statistical equilibrium are

$$p(x + 1) = \frac{A}{(x + 1)} p(x) \qquad (0 \leqslant x < \infty)$$

The solution is $\qquad p(x) = e^{-A} \dfrac{A^x}{x!}$

A call is unsuccessful if the number of calls already in the system is equal to, or greater than, the number of trunks. Hence the probability of blocking is

$$\sum_{x=N}^{\infty} p(x) = e^{-A} \sum_{x=N}^{\infty} \frac{A^x}{x!}$$

This formula gives slightly higher blocking for a given traffic than Erlang's. Thus, at a loss probability of 0·01, Molina's formula allows about 8% lower traffic capacity than the latter. This is sometimes regarded as a useful margin of safety, as it tends to compensate for the effect of departures from statistical equilibrium due to day-to-day variations in busy-hour traffic.[58] This aspect is not overlooked by users of Erlang's formula; it can be allowed for by suitable overload criteria (Section 2.8). Thus, if the formulas are employed on a strictly comparable basis, there is probably little difference in their practical results. Molina's formula has been used chiefly in the USA.

Situations in which the lost-calls-held assumption is realistic, as distinct from merely convenient, are not common, but they do occur in telecommunications and other fields. A possible example is a time-assigned speech-interpolation system in which the talking periods of a call are inserted into the silent periods of other calls, to make greater use of trunk capacity. If all trunks are busy when a talking period begins, the speaker goes on talking; if a trunk is released before he reaches a pause, it is seized and held for the unelapsed part of the talking period. Provided this does not happen too often, the speech remains intelligible, so congestion does not affect the traffic pattern. In this system, it might appear that N trunks can carry more than N erlang since there may be N callers talking simultaneously, while other calls are in silent periods. Strictly speaking, however, the traffic flow should be defined in terms of the actual occupation times of the trunks, i.e. the spurts of continuous speech, not the total call durations, in which case the traffic carried cannot exceed N erlang.

3.8 Frequency and duration of states

In traffic engineering, we are primarily concerned with the probable state of the system at a certain time, usually that of a call arrival, from the subscriber's point of view. The frequency and duration of states are also of interest, however, especially in problems connected with overloading of tone generators etc., where prolonged overloads may cause damage to equipment, and not merely inconvenience to subscribers.

Consider a full-availability group of N trunks with Poisson input of A erlangs, and average holding time T expressed in hours. During a long period of Z hours the number of calls is AZ/T, so the number of occasions when a call arrival changes the state of the system from x to $x + 1$ simultaneous calls is

$$\left(\frac{AZ}{T}\right)\left(\frac{\dfrac{A^x}{x!}}{\displaystyle\sum_{r=0}^{N}\frac{A^r}{r!}}\right)$$

The total time occupied by states containing more than x calls simultaneously is

$$\left(\frac{\displaystyle\sum_{r=x+1}^{N}\frac{A^r}{r!}}{\displaystyle\sum_{r=0}^{N}\frac{A^r}{r!}}\right)Z$$

The average duration of these states is obtained by dividing the first expression into the second, giving

$$\frac{x!T}{A^{x+1}}\sum_{r=x+1}^{N}\frac{A^r}{r!}$$

the average interval between successive occurrences of these states, i.e. successive transitions from x to $x + 1$ calls, is obtained by dividing the first expression into Z, giving

$$\frac{x!T}{A^{x+1}}\sum_{r=0}^{N}\frac{A^r}{r!}$$

So far, we have assumed that the traffic is in a continuous state of equilibrium, hourly and daily variations being ignored. A more realistic estimate of the average interval between peak states can be obtained by dividing the above expression by the proportion of time during which the traffic is at or near the busy-hour level, neglecting the chance of peaks at other times. Alternatively, greater accuracy can be obtained by taking the traffic distribution outside the busy hour into account, as follows:

Suppose a typical week or other convenient period contains Z_1 hours with a traffic flow of A_1 erlangs, Z_2 hours with A_2 erlangs, and so on. The average number of transitions from x to $x + 1$ simultaneous calls is

$$\sum_{s}\left(\frac{Z_sA_s}{T}\right)\left(\frac{\dfrac{A_s^x}{x!}}{\displaystyle\sum^{N}\frac{A_s^r}{r!}}\right)$$

The total time occupied by states with more than x simultaneous calls can be calculated approximately by treating the process as equivalent to a succession of equilibrium processes, with the traffic flow stepping from A_1 to A_2 after Z_1 hours, and so on. The number of busy trunks is assumed to have an Erlang distribution with the mean value appropriate to the time under consideration, giving

$$\sum_s \left(\frac{\sum_{r=x+1}^{N} \dfrac{A_s^r}{r!}}{\sum_{r=0}^{N} \dfrac{A_s^r}{r!}} \right) (Z_s)$$

As before, the average duration of peaks and the average interval between them are calculated by dividing the first expression into the second and into $(Z_1 + Z_2 + ...)$, respectively. By putting $x = N - 1$, it can be seen that the average duration of the congestion state (all trunks busy) is T/N, irrespective of the traffic flow and its variations; this would be expected, since once this state has been established, its duration depends only on the time of the first call termination, and is unaffected by further call arrivals.

If the number of sources M is finite, so that the input is non-Poissonian, the following method can be used. Let A, B denote the events 'a call arrives during a specified short period dt', and 'there are x simultaneous calls at the beginning of dt', respectively. Let $T = 1$ for convenience.

Let $p(A), p(B)$ denote the prior probabilities of A and B, and let $p(A|B), p(B|A)$ denote the conditional probabilities of A given B, and B given A, respectively.

Then
$$p(A) = A dt$$

$$p(B) = \frac{\binom{M}{x} a^x}{\sum_{r=0}^{N} \binom{M}{r} a^r}$$

where a = calling rate per free source (Section 3·5).

$$p(A|B) = (M - x) a dt$$

Thus
$$p(B|A) = \frac{p(B) p(A|B)}{p(A)}$$

$$= \frac{(M - x) \binom{M}{x} a^{x+1}}{A \sum_{r=0}^{N} \binom{M}{r} a^r}$$

where A = total traffic offered.

The number of transitions from x to $x + 1$ simultaneous calls during Z is $AZp(B|A)$; dividing this into the time occupied by states with more than x

simultaneous calls we obtain the average duration of these states, which is

$$\left\{ (M-x)\binom{M}{x}a^{x+1} \right\}^{-1} \sum_{r=x+1}^{N} \binom{M}{r}a^{r}$$

As before, the average duration of a congestion state $(x = N-1)$ is $1/N$.

3.9 Marginal utility

Because of the relative inefficiency of small trunk groups, it is sometimes argued that the usual practice of adopting a fixed grade of service, irrespective of group size, is uneconomical. An alternative approach is based on the principle of equal marginal utility.[74, 75]

Let C be the cost of one additional trunk, and w the value of the traffic per erlang, expressed in the same units. If the number of trunks in a full-availability group is increased from N to $N+1$, the extra traffic carried is, with A erlangs offered,

$$A[1 - E_{N+1}(A)] - A[1 - E_N(A)] = A[E_N(A) - E_{N+1}(A)] \equiv F_N$$

It is worthwhile adding another trunk to the group so long as the value of the additional traffic exceeds the additional cost, i.e.

$$wF_N > C$$

Hence the optimum value of N is given by

$$\frac{wF_N}{C} \leqslant 1 \leqslant \frac{wF_{N-1}}{C}$$

The value of w may be taken as the average revenue per erlang for the group in question, or it may be adjusted to allow for indirect social benefits, depending on the policy objectives of the administration. For a given value of w/C, the method is equivalent to stipulating a constant increase in traffic carried per additional trunk, rather than a constant probability of congestion, for different sized groups. The effect is to increase the loading on small groups. As an example, $3 \cdot 961$ E offered to 10 trunks gives congestion probability $0 \cdot 005$ (Erlang's lost-calls-cleared formula), while

$$F_{10} = 3 \cdot 96[E_{10}(3 \cdot 96) - E_{11}(3 \cdot 96)] = 3 \cdot 96(0 \cdot 0050 - 0 \cdot 0018) = 0 \cdot 0127$$

The traffic capacity (A) of a single-trunk route with the same marginal utility is given by

$$A[E_1(A) - E_2(A)] = 0 \cdot 0127$$

Whence $A = 0 \cdot 132$

The congestion probability is

$$\frac{A}{1+A} = 0 \cdot 117$$

If the total number of trunks to be divided among several groups is fixed, division in accordance with the principle of equal marginal utility gives the maximum total traffic carried and the best average grade of service.[74]

Chapter 4

Lost-call theory for gradings

When the number of trunks in a route exceeds the outlet capacity of the switches giving access to it, so that, in a step-by-step system, a call has access to only a limited number of trunks in the route, *limited availability* is said to exist. It is possible to divide the route into a number of separate trunk groups, each connected to its own switching matrix and fully available to the inlets of that matrix. This, however, is generally an inefficient arrangement, and greater traffic capacity can be obtained by connecting the route to a single 'incomplete' matrix which, in each group of inlets, has access to the same number of outlets, but not exclusively; that is to say, at least some of the outlets accessible to any subset of inlets are also accessible to others. This technique is known as *grading*, and is effected by partial commoning of the switch outlets.

The principle is illustrated in Fig. 22*a* which shows one of many possible gradings for a matrix of eight inlets and six outlets. The outlets are divided between two routes; a full-availability group of two trunks, and a grading of four trunks at availability two. These are shown in simpler form in Fig. 22*b* and *c*; in the latter, each row (or *grading group*) represents two inlets, both connected to the same two outlets. Fig. 22*d* shows an alternative arrangement of route 2, divided into two separate groups, each being fully available to four out of eight inlets. The advantage of grading can be seen by considering a situation when any two rows experience simultaneous peaks, while the other two rows happen to be slack. The number of trunks available to each pair of groups is shown in Table 3.

The busy groups are served by three or four trunks in the grading and by two or four in the ungraded arrangement, the average being the same in both cases. Thus, grading tends to reduce the effect of unbalanced group loading, whether this is due to random fluctuation or unevenly distributed traffic sources.

4.1 O'Dell gradings

The principles of grading design for Strowger exchanges were established through

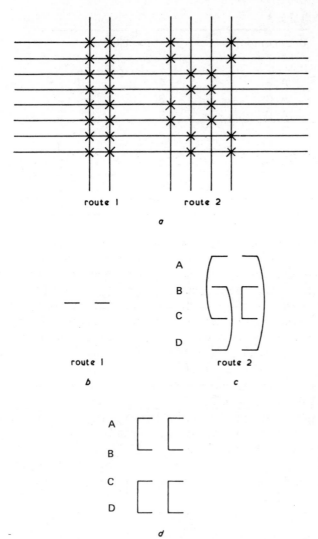

Fig. 22 Principle of grading

an extensive series of simulation tests (Chapter 14) carried out by G.F. O'Dell and others. Fig. 23 illustrates a simple case. It represents a single level of 20 2-motion selectors (Section 1.8) with 10 outlets per level. The selectors are divided into two groups of ten each, and the corresponding outlets of all the selectors in a group are connected together to a distribution frame, where the connection pattern can easily be changed to suit traffic conditions (Fig. 23). Thus, the selectors could be arranged to give access to two separate groups of up to ten trunks each. In this example, however, the traffic is assumed to require only 16 trunks, so the last four outlets

Table 3

Busy groups	Available trunks	
	Fig. 22c	Fig. 22d
A, B	4	2
A, C	3	4
A, D	3	4
B, C	3	4
B, D	3	4
C, D	4	2
Total	20	20
Mean	3·3	3·3

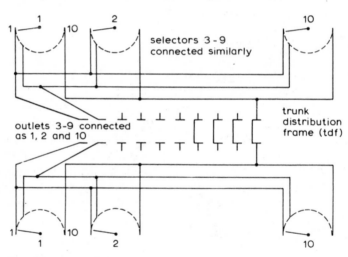

Fig. 23 Grading of Strowger switches

are commoned to form a 2-group grading of availability 10. Any number of groups can be graded in this way, subject to practical limitations to facilitate call tracing and to reduce crosstalk between commoned outlets.

There are usually a number of alternative grading patterns for any particular availability, number of grading groups, and number of trunks. The most suitable gradings for use with Strowger and other switches designed to hunt from a fixed home position are known as *progressive gradings*; that is to say, more grading groups are commoned to each trunk on the later than on the earlier choices. *O'Dell gradings* are of this type, with the further property that only outlets with the same ordinal number in adjacent groups are commoned (Figs. 24*b*–*d*). This was a practical restriction designed to facilitate the maximum use of bare-wire commons, which are cheaper and less liable to faults than covered wiring.

Clearly, the maximum number of trunks which can be served by groups at availability k is gk, arranged as g full-availability groups of k trunks each. If the number of trunks is only slightly less than gk, nearly all trunks are still exclusive to particular groups, so the advantage of grading is not fully exploited. Thus, there is an upper limit to the number of trunks which can be efficiently graded from a given number of groups.

An empirical rule for determining the minimum number of groups for N trunks at availability k is as follows. If $N/K > 3.50$, $g_{min} = 2(N/k)$. Below this range the following values apply:

$$N/k \leqslant 1.60 \qquad g_{min} = 2$$

$$1.60 < N/k \leqslant 2.10 \qquad g_{min} = 3$$

$$2.10 < N/k \leqslant 2.50 \qquad g_{min} = 4$$

$$2.50 < N/k \leqslant 2.86 \qquad g_{min} = 5$$

$$2.86 < N/k \leqslant 3.16 \qquad g_{min} = 6$$

$$3.16 < N/k \leqslant 3.50 \qquad g_{min} = 7$$

Odd numbers of groups are usually avoided, but may be tolerated to avoid unbalanced loading. O'Dell did not give a rule for determining the maximum number of groups, but this is not a serious practical problem. Excessive groups can always be reduced by commoning the outlets on the trunk distribution frame, whereas splitting a permanently wired multiple to form new groups is an expensive operation. There is no point in greatly exceeding the minimum number of groups, except to the extent necessary to ensure balanced loading. Figs. 24b–d show all the possible O'Dell gradings with $g = 4$, $k = |0$ and $N = 20$. Their order of efficiency depends on the traffic. With very high traffic on all groups, two ungraded groups (Fig. 24a) will both make full use of all the choices, so there is little if any advantage in grading. When the traffic load is reduced, the later choices in an ungraded group would not be fully utilised, so it is advantageous to common a few of these over all four groups and split some of the early choices to compensate. There must not be so many commons, however, that the resulting interference between already highly occupied early choices causes increased congestion. In the middle range of traffic, Fig. 24c is the best grading; with very low traffic, however, it is generally advantageous to provide the maximum number of full commons (Fig. 24d). Exact calculation shows that the order of preference is d, c, b below 5 E, c, d, b between 5 E and 12 E, c, b, d between 12 E and 15 E and b, c, d above 15 E.[76] At normal grades of service, the traffic offered lies in the range 5 E to 12 E approximately, so that, for practical purposes, c is the best grading. This is consistent with an empirical rule of O'Dell's, who recommended that the number of choices with each commoning arrangement should be as nearly equal as possible.

The rule can be formulated as follows:

Determine all the ways in which the choices of the grading can be commoned;

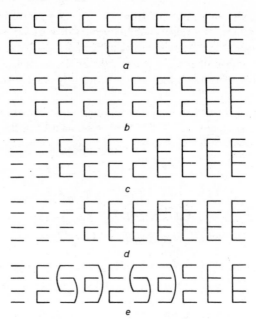

Fig. 24 Progressive gradings

these normally correspond to the different factors of g, which are 4, 2 and 1. Take the differences between the numbers of choices with successive commoning arrangements without regard to sign. Select the grading for which the sum of differences is a minimum.

Thus, in Fig. 24d, the sum of differences is

$$(3-1)+(6-1) = 7$$

In Figs. 23b and c the sum of differences are 11 and 2, respectively, so that c is the best grading.

A more direct method which usually gives a grading with a minimum, or nearly minimum, sum of differences is as follows: this was devised by N.H. Martin (private communication). The principle is to build up the grading, as far as possible, in complete sets of trunks containing each commoning arrangement once. As an example take $k = 10$, $g = 6$, $N = 27$. First, the various arrangements are tabulated, and sets are formed by taking all the arrangements together, then all except the one with the highest number of trunks, and so on, as shown in Table 4.

If the total number of trunks divided by the availability is not less than the average number of trunks per choice for set A, allot choices in arrangement a until the remaining trunks divided by the remaining choices falls below the average trunks per choice for set A. Then allot complete sets A until the average number of remaining trunks per choice falls below that for set B. Then allot complete sets B until it falls below that for set C, and so on. If the total number of trunks divided

Table 4

Arrangement	Trunks from each choice	Set	Arrangements in set	Choices in set	Total trunks in set	Average trunks per choice
a (6 individuals)	6	A	a–d	4	12	3·0
b (3 pairs)	3	B	b–d	3	6	2·0
c (2 threes)	2	C	c–d	2	3	1·5
d (1 full common)	1	D	d	1	1	1·0

by the availability is less than the average number of trunks per choice for set A, but not less than that for set B, omit the first step and begin by allotting set A; if it is less than for set B but not less that for set C, begin by allotting set B, and so on. The same applies at each subsequent step in the calculation. If, however, the number of remaining choices is less than that in the set which should be allocated according to this rule, the set with the right number of choices is allocated.

Thus we have

Required grading:	10 choices,	27 trunks,	2·7 trunks per choice
Provide 1 set A:	4 choices	12 trunks	
leaving	6 choices	15 trunks,	2·5 trunks per choice
Provide 1 set A	4 choices	12 trunks,	
leaving	2 choices	3 trunks,	1·5 trunks per choice
Provide 1 set C	2 choices	3 trunks	
leaving	0	0	

The complete grading comprises two choices of arrangements *a* and *b*, and three each of *c* and *d*.

If the number of trunks remaining after allocating all the choices in accordance with the rules is either more or fewer than required, the difference is adjusted by replacing one or more of the calculated arrangements by others, keeping the sum of differences as low as possible.

Another objective in grading design is to minimise the number of connections which have to be altered when the grading is extended. This may justify minor departures from the minimum-sum-of-difference rule.

4.2 Skipping and slipping

The partial commons in an O'Dell grading are arranged as separate groups, each available to only some of the inlets. Thus, in Figs. 24*b*, *c* and *d*, the upper row of pairs serves the two upper grading groups exclusively. The principle of grading, however, is based on the sharing of different trunks between different sets of inlets, and it is logical to apply this principle to the parts of a grading as well as the whole. Thus, efficiency can be improved, at the cost of a more complex wiring pattern, by 'skipping', i.e. connecting nonadjacent as well as adjacent grading groups in a mixed pattern, so that no two groups are in competition for the same trunks on every paired choice (Fig. 24e). The greater the number of partially commoned choices, the greater, within limits, is the improvement obtainable by skipping; the minimum-sum-of-differences rule does not apply to skipped gradings. It is probably inadvisable, however, to dispense altogether with full commons. Tests on gradings comprising only full commons and pairs, with skipping, indicated a fall in efficiency when the number of full commons was less than five at availability 20, or two at availability 10.[77]

It may be advantageous to use a number of different commoning factors, not so much to increase the efficiency of the grading, but to accommodate a wide range of trunk quantities with a given number of groups in a manner convenient for growth. For example, a grading might eventually require eight groups but initially only four. Eight permanently wired multiples, partially equipped with working circuits may be provided initially. As the traffic increases, additional circuits are equipped in the multiples, and additional outgoing trunks to carry the increased traffic are accommodated by cutting some of the partial commons so as to convert pairs to individuals and fours to pairs.

Tests on 12-group gradings of 93 trunks at availability 20 showed that skipping increased the traffic capacity by about 6% at normal grades of service, and was particularly successful in reducing the effect of unbalanced traffic. In one test, the most heavily loaded grading group carried 30% more than the average traffic per group. With a standard O'Dell grading the blocking was 0·060 on the average and 0·12 for the most heavily loaded group. The corresponding figures with skipping were 0·011 and 0·017.[4]

Another method of commoning is slipping, i.e. interconnection of differently numbered choices of grading groups. This sometimes permits more regular patterns to be formed. As a very simple example, Fig. 25a shows a well balanced 3-trunk, 3-group grading at availability two; without slipping, unbalanced group loading is inevitable (Fig. 25b).

In conjunction with sequential hunting, slipping sometimes introduces additional congestion risks. As an example, compare Figs. 25d and c. Suppose all trunks are free, and a call arrives in group 1, occupying the first choice, and is immediately followed by another call. In Fig. 25d, the second call blocks group 1 if it is in groups 1 or 4, and 2 if it is in group 2, so it has a three in four chance of causing blocking. In Fig. 25c, however, it causes blocking only if it is in groups 1 or 2, the chance being on in two. Thus, other things being equal, a slipped grading may be slightly less efficient than one with skipping between choices of the same ordinal number only. The difference, however, is unlikely to be serious in practical gradings.

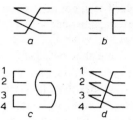

Fig. 25 Skipping and slipping

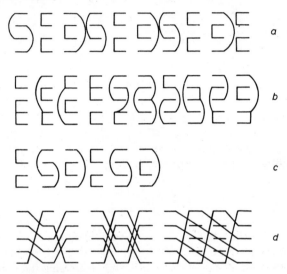

Fig. 26 Gradings with random hunting

4.3 Gradings with nonhoming switches

When nonhoming switches are employed, there is obviously no advantage in differentiating between choices in respect of their commoning patterns. Ideally, the grading should be *homogeneous* (Fig. 26a), i.e. all commons should connect the same number of grading groups (the *interconnection number*). When this is not possible, a *pseudohomogeneous* grading should be formed, using a mixture of two interconnection numbers (Fig. 26b). A homogeneous grading of g groups and of interconnection number c, in which the groups are connected in all $\binom{g}{c}$ possible ways, each combination appearing the same number of times, is said to be *combinatorially homogeneous* (Fig. 26c).

As in the case of progressive gradings, it is often possible to achieve a more regular connection pattern by slipping; this does not affect the traffic capacity. Slipped connections are usually arranged in a cyclic pattern (Fig. 26d). The choices may be divided into a number of sets, not necessarily equal, and all the commons in one set follow a similar pattern; thus, in the left-hand set, the commons connect groups $1-2-5$, $2-3-1$, and so on.

All groups should, as far as possible, be equally loaded. Ideally, it is advantageous, if sufficient groups are available, for each combination of groups to be used once only. The reason is the same as that for preferring progressive gradings with skipping to O'Dell gradings; i.e. to apply the principle of trunk sharing to each part of the grading as well as the whole. In special cases, this can be achieved with a combinatorially homogeneous grading. If k is the availability, g the number of groups and N the number of trunks, the average interconnection number is $c = gk/N$. The number of group combinations, c at a time, is $\binom{g}{c}$, so a complete set of these requires at least

$$\frac{\binom{g}{c}}{g/c} = \binom{g-1}{c-1}$$

choices. A combinatorially homogeneous grading can be formed if

$$k \bigg/ \binom{g-1}{c-1}$$

is an integer; if this is equal to 1, every combination is used once only. The difference in efficiency between gradings with and without repeated combinations may be quite small, so it may not be worthwhile increasing the number of groups merely to avoid repetition. This is illustrated by the following simulation test results for gradings with $N = 40$, $k = 20$, and $g = 4$. The first two have repeated combinations, the second being combinatorially homogeneous. The last two have more combinations than can be used.[90]

Table 5

g	c	$\binom{g}{c}$	Traffic offered, erlangs	Proportion of blocked calls
4	2	6	30·246	0·0332
6	3	20	30·352	0·0282
8	4	70	30·395	0·0256
10	5	252	30·060	0·0232

Assuming optimal design in both cases, sequential hunting is more efficient than random hunting, because late choices are left free as long as possible to cater for peaks. The difference, however, is usually small. As an example, the probabilities of blocking for three gradings are shown below; all have 60 trunks, 6 groups, and availability 20, and the traffic offered is 40 E.

O'Dell grading = 0·0097
Progressive grading with skipped pairs and full commons only = 0·0057
Combinatorially homogeneous grading, interconnection number = 2, random hunting = 0·0075

The first two figures are based on simulation,[77] the latter on interpolation between exact values.[78]

4.4 Traffic capacity of gradings: O'Dell's formula

In principle, the probability of blocking in a grading can be determined from equations of statistical equilibrium, as in the case of a full-availability group. In practical gradings however, the number of distinguishable states, and, therefore, the number of equations to be solved is usually too large for exact solution even with modern computers. Many approximate formulas have been suggested, a few of which are dealt with later.

An important concept in the understanding of the principle of grading is the mutual dependence of two trunks. If they are chosen from widely separated parts of the network, so that common traffic influences are negligible, and each carries a traffic flow of a E, the probability that both trunks are busy simultaneously is a^2. If, on the other hand, they are both part of the same full-availability group, so that the same calls are in competition for both, their states are mutually dependent, and the probability that both trunks are busy is greater. If however, the group is very large, and the trunks are chosen at random, the degree of dependence is small; knowledge of the state of one trunk conveys negligible information about the likely state of the other, or of the group as a whole.

This leads to consideration of Erlang's so-called *ideal grading*. He postulated a grading of N trunks, with a device which, as each call arrives, makes an independent random selection of k trunks. If any of these are free, the call picks up one of them at random; otherwise it is lost. Therefore, k is the availability of the grading. For the present purpose we may assume that N is very large, so that the states of the k choices are practically independent of each other. If A is the total traffic carried in erlangs, each trunk has a probability of A/N of being busy, so the probability that all are busy is

$$B = (A/N)^k$$

Therefore, $$A = NB^{1/k}$$

Thus, for any stipulated probability of blocking B there is a straight-line relation between A and N.

In a practical grading, even if very large, the k links available to a call are not completely independent, since they are permanently associated. However, the fact that they are not all exclusive to any particular subset of inlets does reduce their mutual dependence. The analogy between practical gradings and Erlang's ideal grading was noticed by O'Dell who, like other contemporary investigators, discovered by traffic-simulation tests that efficient gradings likewise exhibited an approximately linear relation between A and N. This led him to suggest the following method of calculating the traffic capacity of a grading.[6] It is assumed that the probability of congestion is low, so that it is not necessary to distinguish between offered and carried traffic.

Let a_k = traffic capacity of a full-availability group of k trunks at a probability of blocking B, calculated from Erlang's lost-call formula.

If $N = k$, the 'grading' reduces to a full-availability group, the average traffic per trunk being a_k/k.

O'Dell assumes that each additional trunk above k carries a_k/k plus a certain proportion q of the additional traffic which it would carry in a large Erlang ideal grading with the same availability and probability of blocking. Then

$$A = a_k + (N-k)\left\{\frac{a_k}{k} + q\left(B^{1/k} - \frac{a_k}{k}\right)\right\}$$

$$= a_k + (N-k)\left\{qB^{1/k} + (1-q)\frac{a_k}{k}\right\}$$

For Poisson input, O'Dell found that $q = 0.53$ and was independent of A and B. Later research has shown, however, that q does vary slightly, and that O'Dell's formula tends to under-estimate traffic capacity under normal conditions by about 5%.[45, 77]

Fig. 27 shows the general form of the N/A curves at availability k and probability of blocking B. Consider a 2-group grading as an example. If all k choices are commoned, it becomes a full-availability group of k trunks corresponding to the point (a_k, k). If the commons are cut on choices with numbers $1, 2, 3 \dots (k-1)$ in succession, we obtain a series of gradings with $k+1, k+2, \dots . (2k-1)$ trunks; cutting the last common giving two separate full-availability groups, represented by the point $(2a_k, 2k)$. The end-points are known from Erlang's formula, and the shape of the intermediate curve is determined by simulation. Similar curves are obtained for gradings of 3, 4 and more groups; in these cases, there may be more than one O'Dell grading for a given number of trunks and groups, so the most efficient grading only is represented on the curve. In the lower part of the curve, this may be a grading with fewer groups than the possible maximum, so that some groups are commoned before grading. It is found that the lower parts of all the curves are approximately coincident and rectilinear; this region represents gradings designed for maximum efficiency in accordance with the rules already explained. When N becomes too large for an efficient grading to be constructed with a given number of

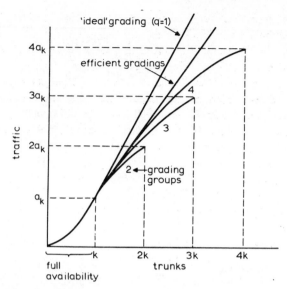

Fig. 27 Traffic capacity of gradings

groups, additional trunks produce a lower increase in traffic capacity, so the curve diverges from the straight line. The curve between $(0, 0)$ and (a_k, k) represents full-availability groups of 1 to k trunks, and is calculated from Erlang's formula.

The line corresponding to $q = 1$ is also shown; this corresponds approximately to Erlang's ideal grading. Practical gradings, with $q < 1$, lie between this line and the line through the points (a_k, k), $(2a_k, 2k)$ etc. It is now known that the 'ideal' grading is not invariably the most efficient possible grading as Erlang thought, at any rate with sequential hunting.[51] For practical purposes, however, it is impossible to obtain very much greater traffic capacity with a given number of trunks, availability and grade of service than that of Erlang's ideal grading, so this arrangement is still a useful standard of comparison.

As an example of the improvement due to grading, the following Table shows the average traffic carried per trunk, at various loss probabilities, by ungraded (full-availability) groups of 20 trunks, and by O'Dell gradings, progressive gradings with good skipping, and Erlang's ideal gradings, all comprising 93 trunks at availability 20. The figures for ungraded groups are based on Erlang's lost-calls-cleared formula; those for the ideal grading are based on the accurate formula for this arrangement, with 93 trunks at availability 20,[142] not O'Dell's adaptation of it to practical gradings with $q = 1$. The other figures are based on the results of simulation tests by Leighton and Kirkby;[4] the skipped gradings are the best of a number of alternative skipping patterns. Poisson input applies in all cases.

4.5 Equivalent random method

This method of calculating congestion in trunk groups which are offered the overflow

Table 6

Probability of loss	ungraded groups	Erlangs carried per trunk		Erlang's ideal grading
		O'Dell grading	skipped grading	
0·001	0·470	0·604	0·655	0·642
0·002	0·502	0·627	0·674	0·668
0·005	0·552	0·663	0·703	0·706
0·01	0·594	0·694	0·731	0·739
0·02	0·647	0·730	0·764	0·775

traffic from other groups was devised by R.I. Wilkinson, primarily for alternative-routing networks (Chapter 11). As it is also applicable to O'Dell gradings, however, it is convenient to treat it in this section. A grading of availability 10, with four groups and 19 trunks, will serve to illustrate the principle (Fig. 28). .

Let a_x, v_x denote the mean and variance, respectively, of the traffic *offered* to each trunk of choice x. In the case of a full-availability group with sequential hunting and Poisson input, the total traffic offered being a_1 erlangs, we have

$$a_x = a_1 E_{1, x-1}(a_1)$$

Also,[7] it can be shown that

$$v_x = a_x \left(1 - a_x + \frac{a_1}{x - a_1 + a_x} \right)$$

In Fig. 28, the traffic offered to each trunk of the third choice comprises the overflows from two 2-trunk full-availability groups; since these are independent, their means and variances are additive. Hence

$$a_3 = 2a_1 E_{1, 2}(a_1)$$

$$v_3 = a_3 \left(1 - \frac{a_3}{2} + \frac{a_1}{3 - a_1 + a_3/2} \right)$$

It has been established by traffic simulation that the probability of congestion in a full-availability group is determined, to a close approximation, by the mean and variance of the traffic offered, irrespective of the sources of that traffic. Thus, for the purpose of calculating the overflow from the fifth choice, we can replace the

Fig. 28 Equivalent random method

first two choices of the two upper grading groups by a single full-availability group of N choices, with Poisson input of Y erlangs; the values of N and Y are chosen so that the overflow from the fictitious group has mean value a_3 and variance v_3. This may require a nonintegral value of N (Appendix 1.8). The fictitious trunks, together with the upper paired choices in the grading, form a full-availability group of $N + 3$ choices. A similar arrangement replaces the first five choices of the two lower-grading groups. The traffic offered to the sixth choice of the grading is equal to the combined overflow from the equivalent groups, its mean and variance being

$$a_6 = 2YE_{N+3}(Y)$$

$$v_6 = a_6\left(1 - \frac{a_6}{2} + \frac{Y}{N+4-Y+a_6/2}\right)$$

Similarly, the first five choices of the whole grading (14 trunks) can be replaced by a full-availability group of N' choices, with Poisson input of Y' erlangs, such that the overflow has a mean value of a_6 and a variance of v_6. This arrangement, together with the five common choices of the grading, forms a full-availability group of $N' + 5$ choices, the overflow from which has mean value

$$a_{11} = Y'E_{N'+5}(Y')$$

The call congestion, which for Poisson input is equal to the time congestion, is the ratio of the overflow to the actual (not fictitious) traffic offered, i.e.

$$\frac{a_{11}}{4a_1}$$

Tests on simple gradings indicated almost exact agreement between the equivalent random method and traffic simulation. Standard curves have been prepared to expedite the calculation of equivalent group sizes and traffic values.[7, 80]

4.6 Palm–Jacobaeus and m.P.J. formulas

Another approximate method of calculating congestion probability for a grading is as follows: Consider a full-availability group of N trunks, with A erlangs offered (Poisson input), and random hunting. If there are x calls in progress, these can be arranged in $\binom{N}{x}$ ways. To determine the probability that a specified set of k trunks ($k \leqslant x$) are all busy, we proceed as follows: If k out of x calls are allocated to the k trunks in question, the remaining calls can be arranged in $\binom{N-k}{x-k}$ ways. Assuming all call arrangements are equally likely, the probability of all k trunks being occupied is the ratio of the number of arrangements of x calls in which these trunks are all occupied, to the total number of possible arrangements of x calls, i.e.

$$\frac{\binom{N-k}{x-k}}{\binom{N}{x}}$$

The total probability that a set of k trunks are all engaged is, since x has an Erlang distribution,

$$B_k = \frac{\sum_{x=k}^{N} \frac{A^x}{x!} \binom{N-k}{x-k} \bigg/ \binom{N}{x}}{\sum_{r=0}^{N} \frac{A^r}{r!}}$$

By rearrangement this becomes

$$\left\{ \frac{\frac{A^N}{N!}}{\sum_{r=0}^{N} \frac{A^r}{r!}} \right\} \left\{ \frac{\sum_{x=k}^{N} \frac{A^{x-k}}{(x-k)!}}{\frac{A^{N-k}}{(N-k)!}} \right\} = \frac{E_N(A)}{E_{N-k}(A)}$$

This formula is usually known as the Palm—Jacobaeus formula, although it was probably first used by T.C. Fry.[52]

Next, consider a grading of N trunks with availability k and A erlangs offered. If the blocking is low, it is not unreasonable to assume that the number of calls in the grading has an Erlang distribution approximately. Each call has access to k trunks, and the probability that these are all busy, which is of course the probability of blocking, is given by the Palm—Jacobaeus formula.

The formula assumes that the traffic carried by a grading is the same as that carried by a full-availability, group with the same offered traffic, whereas it is actually less; hence the formula tends to overestimate the probability of congested states. The error, however, is usually negligible when the blocking is not greater than 0·01. To extend its useful range, it has been modified by replacing the offered traffic A by a value A_0 such that, when A_0 erlangs is offered to a full-availability group of N trunks, the traffic carried Y is the same as that in the grading when offered A erlangs. The probability of blocking in the grading is therefore

$$B_k = \frac{E_N(A_0)}{E_{N-k}(A_0)}$$

This is known as the modified Palm—Jacobaeus (m.P.J.) formula.

We have
$$Y = A_0\{1 - E_N(A_0)\}$$
$$Y = A(1 - B_k)$$

Given N, k and either A or Y, these equations are sufficient to determine B_k, A_0 and Y or A.

Although the m.P.J. formula theoretically assumes random hunting, it has been found to agree with simulation results for progressive gradings with good skipping.[64] In other cases, the following empirical correcting term is added to the m.P.J. traffic capacity.

$$\Delta A = F\left(\frac{N}{k} - 1\right)^2 \frac{k-2}{60+4k}$$

The 'fitting parameter' F is usually negative for practical gradings, which often use simplified connection patterns for convenience.[66]

4.7 Passage probabilities

Although, as already stated, exact solution of the equations of statistical equilibrium for practical gradings is usually impracticable, consideration of the general form of these equations is instructive and leads to some useful approximations. If there are x calls in a grading, the probability that a call terminates during a short interval dt, assuming negative exponential holding-time distribution, is xdt. Assuming Poisson input the probability that a call arrives and is successful during this interval is $Au(x)dt$, where $u(x)$ is the 'passage probability', i.e. the conditional probability that a call is successful, given that there are already x other calls in progress. The equations of equilibrium are as follows: $p(x)$ denotes the probability that the grading contains x calls

$$(x+1)p(x+1) = Au(x)p(x) \quad \{0 \leqslant x \leqslant N, \ p(N+1) = 0\}$$

The probability of blocking is

$$\sum_{x=0}^{N} \{1 - u(x)\}p(x)$$

Clearly $u(x) = 1$ if x is less than the availability k, and $u(N) = 0$. Otherwise, $u(x)$ generally depends on A and on the structure of the grading. Longley has shown that useful information can be derived by considering the limiting values of $u(x)$ as A approaches 0 and ∞.

Consider an O'Dell grading with g groups, N trunks and an availability k, with A erlangs offered. If A is very small, the interference between different groups is negligible, and the probability of any physically possible combination of calls x_1, $x_2 \ldots x_g$ on groups $1, 2, \ldots g$, respectively, is approximately the same as for g full-availability groups of k trunks, with (A/g) erlangs offered to each group. The probability that the grading contains exactly k calls is therefore

$$p(k) = \frac{\displaystyle\sum_{x_1 + x_2 + \ldots + x_g = k} \frac{(A/g)^{x_1}}{x_1!} \frac{(A/g)^{x_2}}{x_2!} \cdots \frac{(A/g)^{x_g}}{x_g!}}{\left(\displaystyle\sum_{r=0}^{k} \frac{(A/g)^r}{r!}\right)^g}$$

$$= \frac{\dfrac{A^k}{k!}}{\left(\displaystyle\sum_{r=0}^{k} \frac{(A/g)^r}{r!}\right)^g}$$

Moreover, with very light traffic, any state which exists is overwhelmingly more likely to have arisen from a lower state by a call arrival than from a higher state by a call termination, since the probability of reaching the higher state is relatively small. It follows that any calls in a grading may be assumed to be on the earliest choices, since gaps can only occur as a result of terminatons. This assumption is sometimes described as 'no holes in the multiple'.[79] Thus, if the grading contains exactly k calls, they cannot cause blocking unless all are in the same grading group; for, if any of the full or partial commons in a blocked group were occupied by calls in other groups, the earlier choices of the latter would also be occupied, and there would be more than k calls in all, contrary to the hypothesis.

The probability that a particular group contains k calls and all the others are empty is

$$v(k) = E_k(A/g) \left(\frac{1}{\displaystyle\sum_{r=0}^{k} \frac{(A/g)^r}{r!}}\right)^{g-1}$$

This expression is the unconditional probability that the grading contains exactly k calls and a particular group is blocked, so we have

$$v(k) = p(k)\{1 - u(k)\}$$

Thus
$$1 - u(k) = \frac{1}{g^k}$$

Other u-terms can be calculated similarly, but unlike $u(k)$ they may depend on the grading pattern..

If A is very great, $p(x-1)$ is negligible compared with $p(x)$; any state which exists is overwhelmingly more likely to have arisen from a higher than from a lower state, and to be ended by an arrival rather than a termination. This leads to a simple limiting expression for $u(n-1)$. Suppose the grading comprises c_r choices with interconnection number r; r is 1 for individuals, 2 for pairs, and so on, and g for full commons. Let $p_r(N-1)$ be the probability of $N-1$ calls, with the idle trunk in an r choice. This state is almost certain to have arisen from the state with N trunks busy, by a call termination on one of the $c_r g/r$ trunks with interconnection number r, and it is almost certain to be destroyed by the arrival of a call from one of the r groups with access to a free tunk. For equilibrium, the rate of occurrence of these two events must balance, so we have

$$\frac{Ar}{g} p_r(N-1) = \frac{c_r g}{r} p(N)$$

Thus
$$p(N-1) = \sum_r p_r(N-1) = \frac{p(N)}{A} \sum_r c_r \left(\frac{g}{r}\right)^2$$

But
$$p(N-1) = \frac{Np(N)}{Au(N-1)}$$

Thus
$$u(N-1) = \frac{N}{\sum_r c_r \left(\frac{g}{r}\right)^2}$$

Other u-terms can be calculated in a similar way. $u(x)$ is always higher for $A \to 0$ than for $A \to \infty$ because of the 'no holes in the multiple' effect. Upper and lower limits for the probability of blocking for any value of A can be obtained by substituting the limiting values of $u(x)$ in the equations of equilibrium. Longley gives the following results for the gradings in Figs. 24b, c and d, with 6·86 E offered in each case.[76]

Table 7

Figure	Lower limit	Upper limit
24d	0·00009	0·0011
24c	0·00017	0·0010
24b	0·00063	0·0013

$u(N-1)$ at $A \to \infty$ is the minimum value of the lowest passage probability. The relative order of efficiency of different O'Dell gradings with the same values of g, N and k, with heavy traffic, can be assessed from this term alone, the grading with the highest $u(N-1)$ being the most efficient.

$u(k)$ at $A \to 0$ is the maximum value of the highest passage probability, and provides useful information on its own. For example, if the first two sets of individuals in Figs. 24c and 24d are removed, the values of $u(k)$ for the remaining parts of the gradings are $1 - (2)^{-8}$ and $1 - (4)^{-8}$, respectively; hence d is the more efficient grading at very low traffic. Similarly, by removing the first choice from Figs. 24b and 24c, it is seen that the latter is more efficient.

Kruithof[78] has applied a similar technique to homogeneous gradings with random hunting.

A simple approximation for passage probabilities is given by the 'geometric group concept'.[65] If there are $N-1$ calls, the chance that the free trunk lies in a particular set of k is k/N. Hence the blocking probability is

$$1 - u(N-1) = 1 - \frac{k}{N}$$

We now assume that

$$1 - u(N-2) = \left(1 - \frac{k}{N}\right)^2$$

$$1 - u(N-3) = \left(1 - \frac{k}{N}\right)^3 \quad \text{etc.}$$

4.8 Gradings with limited sources

By obvious analogy with the Palm–Jacobaeus formula, the following expressions can be derived for the time and call congestion of a grading of N trunks at availability k, with M traffic sources, the calling rate per idle source being a erlang (Section 3.5)

$$\text{time congestion:} \quad E_k = \frac{E_N(M, a)}{E_{N-k}(M-k, a)}$$

$$\text{call congestion:} \quad B_k = \frac{B_N(M, a)}{B_{N-k}(M-k, a)}$$

where

$$E_N(M, a) = \frac{\binom{M}{N} a^N}{\sum_{r=0}^{N} \binom{M}{r} a^r}$$

$$B_N(M, a) = \frac{\binom{M-1}{N} a^N}{\sum_{r=0}^{N} \binom{M-1}{r} a^r}$$

As before, a can be expressed in terms of the offered traffic A as follows:

$$a = \frac{A}{M - A\{1 - B_N(M, a)\}}$$

By analogy with the m.P.J. formula, these formulas can be improved by replacing a

with a_0, which is the calling rate that would produce in full-availability group the same carried traffic as calling rate a does in the grading. The modified formulas have been designated B.Q. formulas ('Bernoulli Quotient').[64] They agree well with simulation results for progressive gradings with good skipping, and can be adapted to simplified gradings in the same way as the m.P.J. formula.

Fig. 29 shows a special type of limited-source grading known as 'transposed multiple'. $n_1 n_2$ sources are graded to $n_1 m_2 + n_2 m_1$ trunks at an availability of $m_1 + m_2$, in such a way that each source shares m_1 trunks with $n_1 - 1$ other sources and m_2 trunks with a different set of $n_2 - 1$ sources.

A congestion formula for this arrangement has been derived by Cohen and Beukelman.[81] It is assumed that a call is equally likely to be offered to either a row or column first. The average traffic offered to each column by one source is

$$b_2 = \frac{y}{2} (1 + p_1)$$

where y = average traffic in erlangs offered per source;
p_1 = call congestion for one row.
Similarly, the traffic offered to each row by one source is

$$b_1 = \frac{y}{2} (1 + p_2)$$

Ignoring the nonPoisson character of traffic overflowing from a row to a column or vice versa, each row is treated as a full-availability group of m_1 trunks with n_1 sources offering $n_1 b_1$ erlang. p_1 is calculated from Engset's formula by iteration. The columns are treated similarly, giving p_2. The total call congestion is $p_1 p_2$.

4.9 Variance of overflow traffic from a grading

To apply the equivalent random method to alternative routing schemes, it is sometimes necessary to known the variance of the traffic overflowing from a grading on to an alternative route. In the case of an O'Dell grading, this can be calculated in the same way as the variance of the traffic offered to each choice in the grading, as already explained (Section 4.5). The following method of calculation is applicable to any type of grading; it is due to A. Lotze.[82, 149]

Consider a full-availability group of k trunks offered A_0 erlang of Poisson traffic, of which R_0 erlang overflows. The variance of the overflow is

$$V_0 = R_0 \left(1 - R_0 + \frac{A_0}{k - A_0 + R_0 + 1} \right)$$

The 'variance coefficient' is defined as

$$D_0 = V_0 - R_0 = R_0^2 \left(\frac{A_0}{R_0(k - A_0 + R_0 + 1)} - 1 \right)$$

$$\equiv R_0^2 p$$

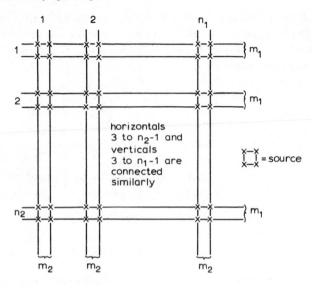

Fig. 29 Transposed multiple
The crosspoints in this diagram connect sources to links, not horizontals to verticals

p is a measure of the 'peakedness' of the overflow traffic; Poisson traffic might be regarded as the 'overflow' from 0 trunks, in which case, putting $k = 0$ and $R_0 = A_0$, we have $p = 0$.

Let us compare the full-availability group with a grading of N trunks at availability k; suppose that the traffic loads offered are such that both have the same probability of congestion. Thus, in any particular grading group, the proportion of time during which all k trunks are busy, during a long period, is the same as for the full-availability group. Moreover, the average duration of each occurrence of this state is the same, i.e. $1/k$ average holding times; hence, the average rate of occurrence of congestion, and the average interval between occurrences, are also the same in both cases. The only difference lies in the distribution of the interval between successive occurrences of congestion. Neglecting this difference, it is reasonable to apply the same value of p in both cases.

Further approximation is necessary when applying this result to calculate the variance of the overflow from the whole grading, because the correlation between different grading groups is difficult to allow for accurately; Lotze gives two approximations, which represent lower and upper limits of D. The first consists in treating the grading as equivalent to N/k independent full-availability groups of k trunks. For simplicity, N/k is assumed to be an integer, but the result is applicable without this restriction.

Let A = total traffic offered to the grading
$\qquad B_k$ = probability of congestion, calculated from the m.P.J. or another appropriate formula
$\qquad A_0$ = traffic which, when offered to a full-availability group of k trunks gives congestion B_k.

Having determined A_0, p can be calculated from the above formula, with $R_0 = A_0 B_k$. The actual overflow from the grading is $R = AB_k$ and we treat this as coming from N/k equally loaded independent groups so that the variance is additive. The variance coefficient is therefore

$$D = \frac{N}{k}p\left(\frac{R}{N/k}\right)^2 = \frac{pR^2 k}{N}$$

and the lower limit of the variance is

$$V = D + R = R(1 + pkR/N)$$

An approximate allowance for the effect of correlation between grading group gives the following expression for the upper limit of D

$$D = pR^2\frac{k}{N}\left(1 + \frac{N-k}{gk}\right)$$

(g being the number of groups).

For practical purposes the arithmetic mean of the upper and lower limits is a good enough approximation.

4.10 Traffic capacity of gradings with 'peaky' input

If the traffic offered to a grading consists wholly or partly of traffic overflowing from other routes, the total variance of the input is the sum of the variances of each component, which can be calculated as already described for full-availability groups and gradings; the case of link systems will be dealt with later (Section 6.5). The following method can then be used to determine the number of trunks in the grading for a given probability of congestion, or vice versa:

Let $\quad k$ = availability
$\quad\quad A, V$ = mean and variance of total traffic offered
$\quad\quad B_k$ = required probability of blocking.

The first step is to determine a fictitious Poisson traffic Y, and a fictitious availability k_0 and number of trunks N_0, such that, when Y erlangs are offered to a grading of N_0 trunks at availability k_0, the overflow traffic has mean value A and variance V; tables are available to facilitate this calculation.[150] The fictitious and

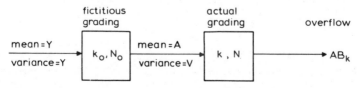

Fig. 30 Grading with 'peaky' input

actual gradings are now combined to form a single grading with Poisson input Y and availability $k_0 + k$ and $N_0 + N$ trunks, the fictitious portion being searched first (Fig. 30). The required overflow is AB_k; this represents a probability of blocking, in the combined grading, of

$$AB_k/Y$$

An appropriate formula, such as the m.P.J. is now used to calculate the number of trunks in the combined grading, giving $N_0 + N$, and then N, the number in the actual grading. Similarly, B_k can be calculated if N is given.

Chapter 5

Lost-call theory for step-by-step systems

In a step-by-step system, a call is liable to encounter congestion at any switching stage, since the selection of a trunk at each stage takes no account of traffic conditions elsewhere. If B_r is the probability that congestion is encountered at stage r, the probability that a call suffers congestion when routed through s stages is

$$1 - (1 - B_1)(1 - B_2) \dots (1 - B_s)$$

If all the B terms are small, this is approximately

$$B_1 + B_2 + \dots + B_s$$

This approximation is usually valid at normal grades of service.

5.1 Independent switching stages

It is obvious that the states of successive stages are interdependent; since, apart from blocked calls, setting-up time etc., they carry practically the same traffic. To a first approximation, however, they can often be treated as independent; in that case, the congestion in each full-availability group or grading can be calculated by means of the appropriate formula, in accordance with Chapters 3 and 4 usually assuming Poisson input. The preceding switching stage only appears in the calculation in as much as its outlet capacity determines the availability of the stage under consideration. As an example, let us take a Strowger exchange comprising 24-outlet subscribers' uniselectors, 200-outlet first selectors, 100-outlet second selectors and final selectors (Fig. 31). Grading exists between each rank of switches, except where the traffic is small enough to obtain full availability. As mentioned before, the congestion on first selectors is often calculated on a lost-call basis, for convenience, the facility of waiting for dial tone being ignored; the availability to be used in this calculation is the number of first selectors accessible from each uniselector, which is 24 if the full capacity is used. Uniselectors with two home positions are

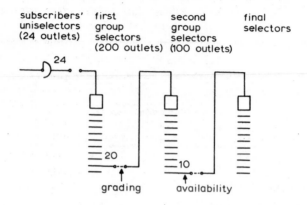

Fig. 31 Diagram illustrating traffic calculations for a step-by-step exchange

sometimes used to guard against the risk of a faulty trunk being picked up repeatedly by a caller, especially during periods of light traffic; this reduces the number of working outlets to 23. These are tested in the order 1–23, or 12–23 and 1–11, on alternate calls. The grading is made up of two progressive gradings, tested in sequence, and each comprising a full range of individual choices, and partial and full commons. The traffic capacity is slightly less than that of a normal progressive grading of availability 23.[45] Similarly, since each first selector level has 20 outlets, the availability of the second selectors is 20, while that of the final selectors is 10. The outlet capacity of the final selectors themselves does not affect their availability to the prefinals. If 200-outlet finals are used, the corresponding outlets from two adjacent prefinal selector levels are connected to the same final selector; a call is routed to the correct half of the final selector multiple according to whether or not the previous digit dialled was odd or even, but the availability is still the number of outlets in one prefinal level. It would be more efficient if the prefinal selector could search two levels in succession when either of the corresponding digits is dialled, thus doubling the availability; this is possible in systems which use large uniselectors instead of 2-motion Strowger selectors.

To calculate switch quantities for a step-by-step exchange a set of traffic capacity tables is required for each availability and grade of service applicable to the system. Appendix 3 shows a table for full-availability groups of 1–100 trunks, based on Erlang's lost-call formula. Assuming that an estimate of the traffic on the route has been obtained from measurements and growth forecasts, the number of trunks required for a specified grade of service can be read from the table. If the traffic lies between two tabulated values, either the next higher or the next lower is chosen, depending on the standard practice of the administration.

5.2 Traffic smoothing[61,154]

If A is the total traffic in erlangs offered to the exchange, the probability of a call being offered to the first rank of switches during a time dt is Adt (Section 2.4), irrespective of the number of existing calls. The probability of a call being offered to a later rank, however, is reduced as the number of existing calls increases, because a higher proportion of calls are lost at an earlier stage, so the input to the second stage and to subsequent stages is not strictly Poissonian. In other words, the effect of interstage dependence, which we have so far neglected, is to reduce the probability of blocking. The magnitude of the error depends on the trunking scheme. In the scheme of Fig. 2, for example, the effect is small; the knowledge that a particular set of 10 A–B links is free from congestion has little bearing on the state of the traffic as a whole, or on the probability of the required set of 10 B–C links being congested. On the other hand, when a high proportion of the traffic entering a rank of switches emerges on one route, the degree of dependence between inlets and outlets is relatively high.

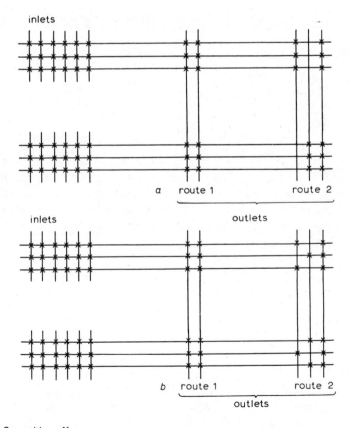

Fig. 32 Smoothing effect

This 'smoothing effect' is increased, in the case of outlets with limited availability, by the usual practice of arranging the interconnections between successive switching stages in such a way that a localised peak at one stage is widely spread over the switch in the next stage. Figs. 32*a* and *b* illustrate this effect in a simple case, which is not, of course, intended as a realistic exchange-trunking scheme. In both arrangements, two groups of inlets are connected to five trunks via two ranks of switches. The number of inlets is assumed to be very large, and will be treated as infinite for the purpose of calculation. The trunks are divided between two routes, comprising a full-availability group of two trunks and a grading of three trunks at availability 2.

Two of the trunks in the grading are each accessible to only three of the six intermediate links, while the third is accessible to all six. In arrangement *a*, however, each of the first two trunks is accessible to only half the inlets, while in arrangement *b* both are accessible to all inlets. As there is no conditional selection, however, it is possible for a call to fail in either arrangement even when one of the partially accessible trunks is free, if the other two trunks are both busy and the call happens to pick up an intermediate link which is not connected to the free trunk. In the first instance, sequential hunting of both ranks is assumed; from top to bottom for intermediate links, and from left to right for trunks in the grading. Let us assume that the system is empty, and that the first three calls are for the graded route. It is impossible for either of the first two calls to be blocked, but, in arrangement *a*, the third call will be blocked if all three are from the same inlet-group and the third arrives before either of the first two ends; in arrangement *b*, however, it cannot be blocked under any circumstances.

If random hunting is used on the intermediate links and sequential hunting is used on the graded trunks, the third call can be blocked in either arrangement, but there is less chance of this happening in *b* than in *a*. To show this, we may assume, without loss of generality, that the first call is from the upper group of outlets. The subsequent sequence of events will be denoted as follows: 2U3L1 means that the first call seizes intermediate link no. 2, counting from the topmost link down; the second call is in the upper inlet-group, and seizes intermediate link no. 3; and the third call is in the lower inlet-group and seizes intermediate link no. 1. The link number refers to the links available to the inlet-group in question; the upper and lower sets of links are both numbered 1-3. In arrangement *b*, the sequences which result in blocking of the third call are as follows: 1U3L2, 1L2U3, 2L1L3, 2L3L1, 3U1L2, 3L2U1. The probability of the first term is always 1/3, and that of the second term is always 1/2. If the second call is in the same inlet-group as the first, the probability that it occupies a particular one of two free links is 1/2; if it is in the other inlet-group, the probability of seizing a particular link is 1/3. The fourth term has probability 1/2, and that of the fifth is either 1/2 or 1/3, depending on whether or not the third call is preceded by another call in the same inlet-group. For example, the probability of the sequence 1U3L2 is $(1/3)(1/2)(1/2)(1/2)(1/3)$. The sum of the probabilities of all blocking sequences turns out to be 1/12. In the case of arrangement a, the third call is blocked if, and only if, all three calls are in the same inlet-group, the probability of which, conditional on the first call being in the upper inlet-group as assumed, is $1/4$.

We have seen that the variance of the traffic offered decreases with the number of calling sources, if the total traffic flow is kept constant, because the probability of a call arrival decreases as the number of busy sources increases (Section 3.5). The smoothing effect of interconnection patterns can be interpreted in an analogous way. The left-hand trunk in the graded route in arrangement b cannot be offered a call by the upper inlet-group when both intermediate links are busy, or by the lower inlet-group when the only intermediate link is busy. In arrangement a, calls are prevented only when all three intermediate links are busy, which is less likely than the afore-mentioned states, so that the smoothing effect is less.

There is obviously less dependence between the total number of calls in the first stage, and those on a particular route from the second stage, than there is between the total calls in both stages. Thus the dependence effect is greatest when all the traffic is on one route; this sometimes occurs in practice when a rank of switches is introduced merely to conform with the numbering scheme, although no division of traffic is required at that stage. The same applies to the smoothing effect due to the interconnection pattern. As the proportion of traffic on a route, and therefore the number of trunks, decreases, so does the number of grading groups over which traffic peaks can be dispersed, until finally there is only a single full-availability group, as in route 1, Fig. 32; in that case, the smoothing effect becomes negligible, and Erlang's lost-call formula is applicable.

The magnitude of the smoothing effect depends on the trunking scheme. A practical check can be carried out as follows: Calculate the number of trunks which would be required to carry the traffic from a rank of switches, on the assumption that it was all carried on one route, using traffic tables for gradings with Poisson input. If these tables call for more trunks than there are switches in the rank, they are obviously inaccurate; for it would be possible to divide the switches into sets of k, where k is the availability of the route, each set being connected to k trunks at full availability, with no blocking. Thus, there is no point in providing more trunks on a route (i.e. one level of a group of Strowger selectors) than the number of switches giving access to them. Otherwise, the smoothing effect is unlikely to be significant, so the Poisson input tables are probably applicable. This is usually the case when the first switch rank is divided into a number of small, lightly loaded groups. On the other hand, if the trunking scheme produces high concentration at the first stage, there may be significant smoothing at later stages. For example, an exchange with 24-outlet subscribers' uniselectors would probably have its first selectors loaded to about 70% to give a grade of service of, say, 0·005, although the average occupancy might be lower if these are in the same multiple as incoming selectors associated with a number of small junction groups. If the first selectors have 100 outlets, giving an availability of 10 per level, then, if all the traffic were on one level, the trunk occupancy, assuming Poisson input, would be about 50%—60% depending on the type of grading. Thus, by the rule given above, smoothing may be significant.

It is clear from the foregoing discussion that the traffic capacity of the second and later switch ranks may depend on a number of parameters, including the occupancy of the preceding switches and the proportion of traffic carried on each route. The effect of smoothing in a Strowger system, with a 2-stage uniselector link system giving access to the first selectors, was investigated experimentally by Grinsted, Dumjohn and Martin,[84] and the results were used to compile traffic tables taking account of these parameters. O'Dell adopted a less accurate method, designed to cater for average conditions of smoothing; he used the same formula as that for gradings with Poisson input, but with the empirical constant $q = 1$ instead of 0·53. In other words, he found that, with smooth traffic, an O'Dell grading could be as efficient as Erlang's ideal grading with the same availability and number of trunks, although this was admittedly a coincidence. In using O'Dell's tables, care must be taken that the number of trunks on any route does not exceed the total number of switches in the preceding rank, which, as we have seen, is unacceptable. In practice, the smoothing effect is not usually very important beyond the second switching stage. O'Dell laid down a number of empirical rules, which will not be quoted here, to ensure that the smooth traffic tables were only used when the magnitude of the effect justified it.

5.3 Availability

Increasing the availability of a grading, like that of a full-availability group (Chapter 3) improves the traffic capacity, but the rate of improvement falls off so that eventually the cost of a further increase outweighs the benefit. In the case of a Strowger 2-motion selector, the availability is determined by the total outlet capacity since all levels have the same number of outlets. Based on the cost of equipment within the exchange, it has usually been found that the optimum availability is in the neighbourhood of 20 for large exchanges, so that 200 outlets is a convenient standard selector size.

Other types of switch permit the outlets to be divided between different routes in accordance with the traffic and trunk costs. This flexibility, including the elimination of wasted outlets corresponding to unused digits, may produce substantial savings, typically 15% for 100-outlet selectors and 8% for 200-outlets.[50] A simple empirical rule which is probably near enough to the optimum for practical purposes in most cases is to divide 70% of the outlets equally between the routes, and divide the remaining 30% in proportion to the traffic on each route. Trunk costs do not have much effect on the optimum, unless the average availability per group is small (say less than 20) or if any routes carrying a substantial proportion of traffic differ in cost per trunk by a ratio of 3:1 or more. Where necessary, costs can be allowed for by multiplying the traffic on each route by the cost per trunk expressed in terms of any convenient unit, and treating this cost-loaded traffic as if it were the actual traffic for the purpose of calculating availability.

Chapter 6

Lost-call theory: link systems

The earliest theoretical analysis of a system with conditional selection was due to
C. McHenry,[10] who investigated a 2-stage uniselector scheme of the type illustrated
in Fig. 7. This system was also investigated experimentally by F.P. Dumjohn and
N.H. Martin.[11] The widespread adoption of the conditional-selection principle in
modern telecommunication systems led to a need for general methods of congestion calculation applicable to a wide range of trunking schemes. As in the case
of gradings, exact solution is only feasible in simple cases. Quite a number of
approximate formulas have been developed[63] many of which have been proved, by
simulation, to be reasonably accurate, at any rate under the conditions for which
they were originally devised. In fact, considerable simplifications can frequently
be made with negligible effect on numerical results, depending on the degree of
accuracy required. A simple method of calculation may be sufficiently accurate
to enable a system designer to choose between alternative trunking schemes, or to
ensure that the system is sufficiently flexible to meet unusual traffic conditions.
On the other hand, it might be worth using a more accurate method for the purpose
of equipment provisioning, since the resultant saving in costs over a large network,
although a small percentage of the total, might be more than sufficient to justify
the cost of computation.

A few of the most generally useful formulas are discussed later. In view of the
great variety of trunking schemes employed in telecommunication networks, it is
difficult to generalise about the accuracy and range of application of different
formulas. Successful approximation is largely a matter of assessing which congestion states are most significant in a particular system, and making appropriate
simplifications. For example, if the trunking is such that congestion occurs most
frequently through mismatching of free links in different switching stages, while
it is extremely rare for a high proportion of links to be occupied in any one stage,
there is no need to use a formula which gives the tail of the call distribution at one
stage with great accuracy.

6.1 Jacobaeus' combinatorial method

The simple trunking scheme of Fig. 3 will serve to illustrate the principle of this method.[8] The problem is to determine the probability of blocking when an attempt is made to connect a particular A-switch inlet (calling-subscriber's line) to a particular C-switch outlet (called-subscriber's line). The probability of the required outlet being already busy is not included in the congestion, so we need only consider the connection of a particular A-switch inlet to any one of a group of B-switch outlets serving the required C-switch; these will be referred to as the 'route'. In other problems, the route might be a group of junctions to a distant exchange, the switches therein forming part of a separate link or step-by-step system. For simplicity, it is assumed that the number of trunks in the required route is the same as the number of B-switches, as is the case in Fig. 3, so that each B-switch serves one link. As there are only two stages of link selection, this arrangement is described as a 2-stage link system, although there are actually three switching stages.

Fig. 33 shows the same trunking scheme in a form which is more convenient for traffic calculation. Each A column, B row or C column represents the *inlets* of an A, B or C switch. The strokes represent the direction of connection. Thus, each A device has access to all B devices in the same column, and each B device has access to all C devices in the same row. In this example, $m = 10$ and $n = 100$. The function of the network is to connect a particular A device to any C device in a particular column. The diagram can be further simplified by showing only the devices which can be used in a particular connection, the strokes being omitted (Fig. 34). To

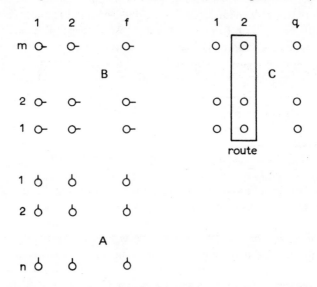

Fig. 33 Diagram for traffic calculation in 2-stage link system

enable a connection to be made, there must be at least one free B device in the same column as the calling A device. A general expression for the probability of congestion is as follows:

$$E_L = \sum_{p=0}^{m} H(m-p)\,G(p)$$

where $G(p)$ = probability that p devices in the C column are busy, $H(m-p)$ = probability that the $m-p$ B devices in the same horizontal rows as the free C devices are all busy.

By definition $H(0) = 1$. Alternatively, $G(p)$ could refer to the B column and $H(m-p)$ to the C column.

Assuming random choice of links we have

$$H(m-p) = \sum_{x=m-p}^{m} J(x)\frac{D(m-p,x)}{C(m,x)}$$

where

$$J(x) = \text{probability of } x \text{ calls in the B column.}$$

$$C(m,x) = \text{total number of ways of arranging } x \text{ calls on } m \text{ trunks.}$$

$$D(m-p,x) = \text{number of ways of arranging } x \text{ calls on } m \text{ trunks, in such a way that a particular set of } m-p \text{ trunks are all occupied.}$$

Then

$$C(m,x) = \binom{m}{x} = \frac{m!}{x!(m-x)!}$$

and

$$D(m-p,x) = \text{number of ways of arranging } x-(m-p) \text{ calls on } m-(m-p)$$
$$\text{trunks}$$

$$= \binom{m-(m-p)}{x-(m-p)} = \binom{p}{x-m+p}$$

Thus,

$$\frac{D(m-p,x)}{C(m,x)} = \frac{p!x!}{(x-m+p)!m!}$$

Jacobaeus showed that reasonable approximations for the G and J functions can be obtained by assuming Bernoulli or Erlang distribution, depending on the trunking scheme, The correlation between traffic at different stages is neglected.

Bernoulli distribution is applicable to a link group serving a relatively small number of sources. Engset distribution would be a better approximation, but is less easy to handle mathematically. Erlang distribution generally applies when the

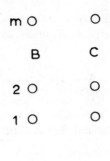

Fig. 34 2-stage link system; 1 B-column and 1 C-column

number of sources is relatively large (say, greater than 30). The sources are assumed to be independent. The validity of this assumption in respect of primary sources, such as subscribers' lines, is dealt with in Section 2.4. From the point of view of the link system, however, some of the immediate sources of traffic may be junctions from other exchanges. As far as possible, each group of junctions is spread over different A-switches, so that the inlets of each A-switch can still be regarded as independent sources, particularly if the junctions are mixed with subscribers' lines. Even if an A-switch serves a small number of junctions from a single large group, these may be regarded as relatively independent, as we have seen in connection with Erlang's ideal grading (Section 4.4). If, however, there is any doubt as to the conditions of independence being met, it may be safer to assume Erlang distribution, which gives greater congestion.

Comparatively simple expressions are obtained for the probability of congestion in the case of 2-stage link systems, as the following examples illustrate:

One B column, Bernoulli distribution; one C column, Erlang distribution (Fig. 34)
Assume in the first instance that there is no concentration at the A stage ($n \leqslant m$). This is a likely case when the A-switch inlets are heavily loaded junctions.
Let

a = traffic carried by each A device, in erlangs

C = traffic offered to the route (C column) in erlangs.

It is assumed that the total number of sources served by the route is large, so that Erlang distribution applies to the C stage.
Then

$$G(p) = \frac{\dfrac{C^p}{p!}}{\displaystyle\sum_{r=0}^{m} \frac{C^r}{r!}}$$

$$J(x) = \binom{n}{x} a^x (1-a)^{n-x}$$

Thus,

$$H(m-p) = \sum_{x=m-p}^{n} \frac{n!\,p!\,x!\,a^x(1-a)^{n-x}}{x!\,(n-x)!\,(x-m+p)!\,m!}$$

$$= \frac{a^{m-p}p!\,n!}{m!\,(n-m+p)!} \sum_{x=m-p}^{n} \frac{(n-m+p)!\,a^{x-m+p}(1-a)^{n-x}}{(n-x)!(x-m+p)!}$$

$$= \frac{a^{m-p}p!\,n!}{m!\,(n-m+p)!}$$

since the series is the binomial expansion of $\left\{ a + (1-a) \right\}^{n-m+p}$.

The time congestion is

$$E_L = \sum_{p=0}^{m} \left\{ \frac{a^{m-p}p!\,n!}{m!\,(n-m+p)!} \right\} \left\{ \frac{(C^p/p!)}{\sum_{r=0}^{m} C^r/r!} \right\}$$

$$= \frac{\dfrac{C^m}{m!} \displaystyle\sum_{p=0}^{m} \dfrac{(C/a)^{n-m+p}}{(n-m+p)!}}{\displaystyle\sum_{r=0}^{m} \dfrac{C^r}{r!} \dfrac{(C/a)^n}{n!}}$$

Since $n \leqslant m$, $(n-m+p)! = \infty$ if $p < m-n$, so the effective range of p is $m-n$ to m, and the right-hand expression is the reciprocal of Erlang's lost-call congestion formula with (C/a) erlang offered to n trunks.

$$E_L = \frac{E_m(C)}{E_n(C/a)}$$

To determine the call congestion, we note that, at the time at which a free A device offers a call, the probability that x of the other A devices in the same column are busy is

$$\binom{n-1}{x} a^x (1-a)^{n-1-x}$$

It follows that the call congestion is obtained by substituting $n-1$ for n in the expression for time congestion, giving

$$E_L = \frac{E_m(C)}{E_{n-1}(C/a)}$$

The same formulas are approximately correct when n is greater than m, but not so large as to justify assuming Erlang rather than Bernoulli distribution at the B stage.

It will be noted that a is traffic *carried* and C is traffic *offered*. At normal grades of service the difference can often be neglected. When this is not the case, as under heavy-overload conditions, the congestion corresponding to known values of traffic offered can be calculated as follows: Let B and C denote the traffic offered to a B and C column, respectively. First, calculate the call congestion B_L with a guessed value of a, the average traffic carried per A device. If $B(1 - B_L) \neq an$, repeat the calculation, using a new value of a, thus obtaining a new value of B_L. The process is continued until successive value of B_L are in sufficiently close agreement. The same value of C is used in each calculation.

A still better estimate can be obtained by starting with assumed values a and c of the traffic carried per A and C devices, respectively; these are chosen so that the total traffic carried by all devices in both stages, not only the routes under consideration, is the same, after allowing for any differences in holding time if the operation of the system requires this. In the Erlang terms, C is replaced by C_0 the traffic which, if offered to the C column at full availability, would produce a carried traffic of c erlang per device; C_0 can be found from tables of the m.P.J. formula[9] (Section 4.6).

One B column, Bernoulli distribution; q C columns, Erlang distribution (Fig. 35)
In this scheme, each B switch is connected to q C devices in the same route. An example is the double uniselector scheme of Fig. 7, where there is only one route and $q = 5$. It is possible to reduce the congestion by reserving one C column, as far as possible, until the others are full. For the purpose of calculation, it is assumed that this is done so effectively that the last C column is always empty unless there are more than $(q - 1)m$ calls in the route; this cannot quite be achieved in practice because call terminations cannot be controlled.

It is assumed, in the first instance, that $n = m$, so that $a = b$. Assuming random occupation of B devices, the probability that a particular set of x of these are all busy is

$$\sum_{r=x}^{n} \binom{n}{r} a^r (1-a)^{n-r} \binom{n-x}{r-x} \Big/ \binom{n}{r}$$

$$= a^x \sum_{r=x}^{n} \binom{n-x}{r-x} a^{r-x}(1-a)^{n-r} = a^x = b^x$$

In other words the B devices, like the A devices, behave as independent sources. If there are fewer than $(q - 1)m$ calls in the route, congestion can exist only if all the B devices are busy. Hence

$$\hat{H}(m-p) = b^m \text{ if } p < (q-1)m$$

$$H(m-p) = b^{mq-p} \text{ if } p \geqslant (q-1)m$$

Thus,
$$E_L = \frac{\displaystyle\sum_{p=(q-1)m}^{mq} b^{mq-p}\frac{C^p}{p!}}{\displaystyle\sum_{r=0}^{mq}\frac{C^r}{r!}} + \frac{b^m \displaystyle\sum_{p=0}^{m(q-1)-1}\frac{C^p}{p!}}{\displaystyle\sum_{r=0}^{mq}\frac{C^r}{r!}}$$

A simple approximation is to extend the first series over the range $0-mq$, and neglect the second, giving, after rearrangement,

$$E_L = \frac{E_{mq}(C)}{E_{mq}(C/b)}$$

which can also be written in the form

$$E_L = \frac{E_{mq}(C)}{E_{nq}(C/a)}$$

This formula is approximately correct even if $n \neq m$, provided n is small enough to justify the assumption of a Bernoulli distribution. The call congestion is

$$B_L = \frac{E_{mq}(C)}{E_{(n-1)q}(C/a)}$$

f B columns, Bernoulli distributions; q C columns, Erlang distribution (Fig. 36)
In this scheme, an A-switch has f links to each B-switch. The Bernoulli stage is likely to have less influence on congestion than the Erlang stage. As an approximation, the effect of concentration or expansion at the B stage will be ignored, and likewise the difference between time and call congestion. Thus, $n = mf$, $a = b$, and the B devices will be treated as independent sources.

In the first instance, we will assume $q = 1$. Then

$$E_L = \sum_{p=0}^{m} b^{f(m-p)}\frac{\dfrac{C^p}{p!}}{\displaystyle\sum_{r=0}^{m}\frac{C^r}{r!}} = \frac{E_m(C)}{E_m(C/b^f)}$$

after rearrangement

If $q > 1$, the following approximation is obtained, by a method similar to that of the preceding case:

$$E_L = \frac{E_{mq}(C)}{E_m(C/b^f)}$$

With concentration in the A stage, these formulas slightly underestimate the congestion. The error is usually small, but, in exceptional cases, may give a lower value for the total congestion than that due to the B stage alone. This anomaly can be avoided by adding Engset's expression for the blocking on a full-availability group of mf trunks, with n sources offering na erlang (strictly $na/(1 - $ call congestion) (Section 3.5)).

1 B column, 1 C column, both with Erlang distribution (Fig.34)
If n is relatively large (say, greater than 30) Erlang distribution may apply to the B as well as to the C stage. Let B and C denote the total traffic offered to the B and C columns, respectively, in erlangs. Then

$$H(m-p) = \frac{\sum\limits_{x=m-p}^{m} \dfrac{B^x}{x!}}{\sum\limits_{r=0}^{m} \dfrac{B^r}{r!}} \left\{ \frac{p!\,x!}{(x-m+p)!\,m!} \right\}$$

$$= \left\{ \frac{\dfrac{B^m}{m!}}{\sum\limits_{r=0}^{m} \dfrac{B^r}{r!}} \right\} \left\{ \frac{\sum\limits_{x=m-p}^{m} \dfrac{B^{x-m+p}}{(x-m+p)}}{\dfrac{B^p}{p!}} \right\}$$

$$= \frac{E_m(B)}{E_p(B)}$$

Thus,

$$E_L = \left\{ \frac{E_m(B)}{\sum\limits_{r=0}^{m} \dfrac{C^r}{r!}} \right\} \left\{ \sum\limits_{p=0}^{m} \frac{C^p}{p!\,E_p(B)} \right\}$$

The right-hand series can be expressed in the form

$$\sum\limits_{r=0}^{m} \frac{B^r}{r!} \sum\limits_{p=r}^{m} \left(\frac{C}{B} \right)^p$$

$$= \left(\frac{B}{B-C} \right) \left\{ \sum\limits_{r=0}^{m} \frac{C^r}{r!} - \left(\frac{C}{B} \right)^{m+1} \sum\limits_{r=0}^{m} \frac{B^r}{r!} \right\}$$

provided $B \neq C$.
Thus,

$$E_L = \frac{BE_m(B)}{B-C} \left\{ 1 - \left(\frac{C}{B} \right) \left(\frac{\dfrac{C^m}{m!}}{\sum\limits_{r=0}^{m} \dfrac{C^r}{r!}} \right) \left(\frac{\sum\limits_{r=0}^{m} \dfrac{B^r}{r!}}{\dfrac{B^m}{m!}} \right) \right\}$$

$$= \frac{BE_m(B) - CE_m(C)}{B - C}$$

If $B = C$ we have

$$\sum_{r=0}^{m} \frac{B^r}{r!} \sum_{p=r}^{m} \left(\frac{C}{B}\right)^p = \sum_{r=0}^{m} (m - r + 1)\frac{B^r}{r!}$$

$$= (m + 1) \sum_{r=0}^{m} \frac{C^r}{r!} - C \sum_{r=1}^{m} \frac{C^{r-1}}{(r - 1)!}$$

Thus

$$E_L = E_m(C)\left\{ m + 1 - \frac{mE_m(C)}{E_{m-1}(C)} \right\}$$

$$= E_m(C)(m + 1 - C) \text{ approximately}$$

Time and call congestion are identical in this case, since the probability of call arrival is independent of the state of the system.

f B columns, 1 C column, both with Erlang distribution
An approximate expression for congestion when there is more than one B column is as follows:[8]

$$E_L \doteqdot \frac{(B/f)\, E_{mf}(B)}{\dfrac{B}{f} - C} + E_m(C) \quad (B/f \neq C)$$

or

$$E_L \doteqdot (m + 1 - C)E_{mf}(B) + E_m(C) \ (B/f = C)$$

This formula is based on the assumption that one B column is empty as long as fewer than $m(f - 1)$ B devices are busy, the calls being packed first into the other columns.

1 B column, 1 C column, both with Bernoulli distribution (Fig. 34)
Bernoulli distribution may be applicable to the C column when it serves a limited number of subscribers' lines. It is convenient to refer to these as 'sinks' in relation to the network from which they receive calls, although most lines, of course, both make and receive calls, thus acting both as sources and sinks. In Fig. 33, the B column serves n sources and the C column serves k sinks.(not shown). Let a and d denote the average traffic carried per source and sink, respectively. Attempted calls to busy sinks have very short holding times, which will be assumed to be zero for the purpose of calculation. The distribution of calls in the C stage is therefore mainly determined by that of the successful calls, on which the limited number of sinks, like that of sources, has a smoothing effect. We will assume, in the first instance, that n and k are both less than or equal to m.

Then

$$G(p) = \binom{k}{p} d^p (1-d)^{k-p}$$

As before,

$$H(m-p) = \frac{a^{m-p} p!\, n!}{m!\, (n-m+p)!}$$

(congestion is possible only if $k + n \geqslant m$)
The time congestion is

$$E_L = \sum_{p=m-n}^{k} \frac{k!\, n!\, d^p (1-d)^{k-p}\, a^{m-p}}{(k-p)!\, m!\, (n-m+p)!}$$

If $k = m$ then $d = c$ and this expression simplifies to)

$$E_L = c^{m-n}(c + a - ac)^n$$

The call congestion is obtained by substituting $n - 1$ and $k - 1$ for n and k.

f B columns, q C columns, both with Bernoulli distribution (Fig. 36)
We will consider only the case with no expansion or concentration, so that $n = mf$, $k = mq$, $a = b$ and $c = d$. The calls on the B and C links then have the same distribution as those on the sources (A) and sinks, respectively, so that the B and C devices can be treated as independent sources. The time congestion is

$$E_L = (b^f + c^q - b^f c^q)^m$$

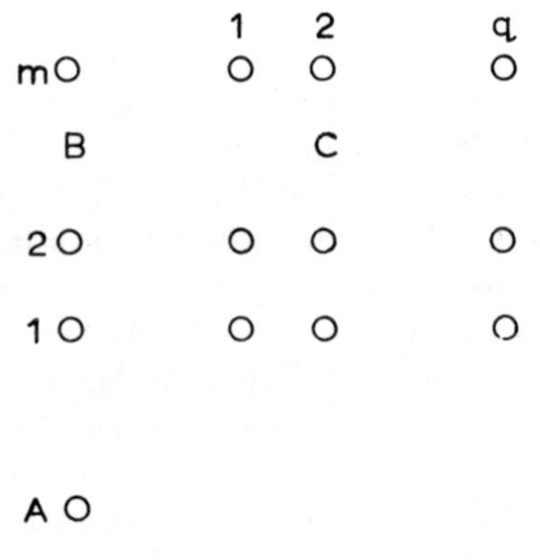

		1	2		q
mO		O	O		O
B			C		
2O		O	O		O
1O		O	O		O
A O					

Fig. 35 2-stage link system; 1 B-column and 1 C-column

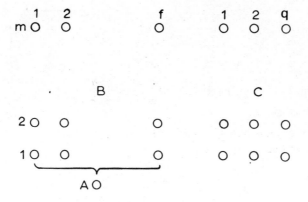

Fig. 36 2-stage link system; *f* B-columns, *q* C-columns

Link systems with grading

In Figs. 33–36, each B column serves a certain set of A devices exclusively, and each A device has access to every C device in the route. This arrangement is analogous to full-availability, although not equivalent to it, since the availability of the route fluctuates with the traffic state of the B stage.

Some link systems, however, employ grading between switch ranks in order to utilise crosspoints more efficiently. Since the conditional-selection principle itself has the same purpose, the improvement obtainable by grading is less than that obtainable in a step-by-step system, but it may be sufficient to justify the wiring complications. It is particularly advantageous to grade the outlets of the first switching stage so as to reduce the effect of unequal calling rates, which inevitably cause traffic unbalance. One method, the 'transposed multiple', is shown in Fig. 29; Fig. 37 shows how such an arrangement can be incorporated in a link system. In this case, $n_1 = n_2 = n$, and $m_1 = m_2 = m$. Each A device has access to m B devices vertically and m horizontally, and each B device has access to one C device in the required route. The distribution in the A and B stages may be assumed to be Bernoulli. For the purpose of calculation, the scheme can be regarded as equivalent to that of Figs. 33 and 34, with m replaced by $2m$. As a margin of safety, the probability of congestion for the transposed multiple alone, calculated from the formula of Cohen and Beukelman, may be added (Section 4.8). The difference between time and call congestion may be neglected.

Fig. 38 shows an arrangement in which each of four A and B columns has access to a different but overlapping set of C-switches. If there were no overlapping, there would in effect be four separate link systems with one B column and one C column each. As it is, the C column is analogous to a grading of n trunks with a maximum availability of m. Jacobaeus analyses this scheme by analogy with O'Dell's theory for single-stage gradings. An alternative approach is as follows: Assume Bernoulli distribution in the B-column, and apply the Palm–Jacobaeus formula to the accessible C devices. The probability of congestion is

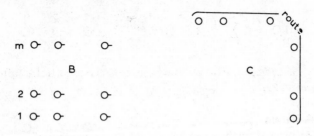

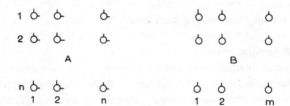

Fig. 37 Link system with transposed multiple

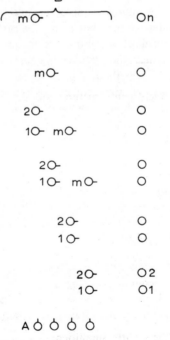

Fig. 38 Link system with grading

$$E_L = \sum_{x=0}^{m} \binom{m}{x} b^x (1-b)^{m-x} \frac{E_n(C)}{E_{n-m+x}(C)}$$

As usual, b is the average traffic per B device; C is the total traffic offered to the whole route (n trunks).

Multistage link system

In principle, Jacobaeus' method can be applied to any number of switching stages, but the number of combinations, and therefore the complexity of the calculation, increases rapidly. The easiest distribution to handle is usually the Bernoulli; in practice, this often gives reasonably accurate results,[85] although it may be necessary to add a correcting term to the total congestion to allow for cases where significant blocking may occur due to high link occupation at one stage as distinct from combinations of states at different stages. Particularly simple expressions may be obtained in the case of series–parallel networks. Fig. 4b shows the paths between one inlet and one outlet in a network of this type (Section 1.3). There are three switching stages, the occupancy of the A-B, B-C and C-D links being a, b and c, respectively. To make connection via the upper A-B link, it is necessary that this link itself, the corresponding C-D link, and the B-C link connecting them, should be free. The probability of this is $(1.-a)(1-c)(1-b)$. Hence, the probability that connection cannot be made via any of three A-B links is

$$E_L = \{1 - (1-a)(1-c)(1-b)\}^3$$

6.2 Dependence between stages

The sources of error in Jacobaeus's method are the assumptions of simplified distributions at each stage and of independence between stages. The effect of dependence is to reduce the probability of congestion. Roughly speaking, this is because peaks at different stages tend to coincide more often than would be expected if they were independent events. Calls are more likely to be lost when peaks exist in at least one stage, therefore overlapping of peaks tends to reduce the total time during which blocking is most likely to occur. It is generally found, in fact, that the combined effect of the two errors is to overestimate the probability of congestion. The magnitude of the error varies considerably according to the trunking scheme and traffic conditions, but, in the author's experience, is unlikely to exceed a factor of 2, which generally corresponds to overprovision of trunks of the order of 5–10%. The error can be practically eliminated by recursive modification of the postulated distributions, stage by stage, at any rate with random link allocation (Section 6.7). To illustrate the principle, we will consider a 2-stage link system as in Fig. 34, which has been investigated by Elldin.[14] The following description of the method is slightly simplified:

Let

a = traffic offered by each free source (Section 3.5)

C = total traffic offered to each C column

$p_B(x)$ = probability that a particular B column contains x calls

$p_C(y)$ = probability that a particular C column contains y calls

$u_B(x)$ = probability of a free path existing between a particular pair of B and C columns, given that the B column contains x calls

$u_C(y)$ = probability of a free path existing between a particular pair of B and C columns, given that the C column contains y calls

n = number of sources served by a B column

m = number of links in a B or C column

ϕ = proportion of traffic from a particular B column destined for a particular C column.

θ = proportion of traffic to a particular C column which comes from a particular B column.

If there are x calls in a B column, the probability that z of them are routed to a particular C column is

$$\binom{x}{z} \phi^z (1 - \phi)^{x-z}$$

If there are at least z calls in the C column, the probability that it contains exactly y calls in all is

$$p_C(y) / \sum_{r=z}^{m} p_C(r)$$

The number of ways in which $y - z$ calls can be arranged in the C column so as to block the $m - x$ free links in the B column, assuming $y - z \geqslant m - x$, is the number of ways in which the remaining calls can be arranged, after $m - x$ of them have been placed in blocking positions; this is

$$\binom{x - z}{y - z - m + x}$$

The total number of ways in which $y - z$ calls can be arranged on $m - z$ links is

$$\binom{m - z}{y - z}$$

The ratio of the first expression to the second is the probability of blocking with particular values of x, y and z. For convenience, the links occupied in the B column are regarded as fixed; further arrangements could be obtained by permuting them, but these would be represented by a multiplying factor of both expressions which would cancel out. It follows that

$$1 - u_B(x) = \sum_{z=0}^{x} \binom{x}{z} \phi^z (1-\phi)^{x-z} \frac{\sum_{y=z}^{m} p_C(y) \binom{x-z}{y-z-m+x}}{\sum_{r=z}^{m} p_C(r)} \binom{m-z}{y-z}^{-1}$$

Similarly

$$1 - u_C(y) = \sum_{z=0}^{y} \binom{y}{z} \theta^z (1-\theta)^{y-z} \frac{\sum_{x=z}^{m} p_B(x) \binom{y-z}{x-z-m+y}}{\sum_{r=z}^{m} p_B(r)} \binom{m-z}{x-z}^{-1}$$

Fig. 39(a) shows a possible blocking arrangement; the filled-in circles represent occupied links, and B and C links occupied by the same call are joined by straight lines.

If a B column contains x calls, the probability of a call arrival during a short time dt is

$$(n - x)a dt.$$

The equations of statistical equilibrium for the B column are therefore

$$(x + 1) p_B(x + 1) = (n - x) a u_B(x) p_B(x)$$

In the case of the C column, we will assume that the total number of sources,

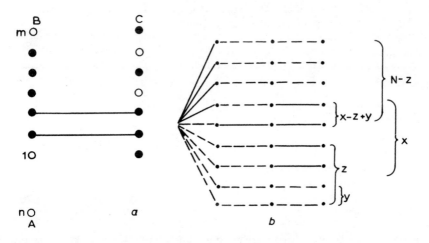

Fig. 39 Diagrams illustrating iterative calculations for link systems

including those served by other B columns, is so large that the probability of a call arrival is $C dt$, independent of the number of existing calls, therefore the equations are

$$(y + 1) p_C(y + 1) = Cu_C(y) p_C(y)$$

For simplicity, it is assumed that u_B and u_C are the same for all pairs of B and C columns.

The procedure is as follows:

(i) Calculate the $p_B(x)$ and $p_C(y)$ terms from assumed distributions; e.g. Engset for stage B, Erlang for stage C.
(ii) Calculate the $u_B(x)$ terms.
(iii) Calculate improved values of the $p_B(x)$ terms from the equations of statistical equilibrium for the B stage.
(iv) Calculate the $u_C(y)$ terms.
(v) Calculate improved values of the $p_C(y)$ terms from the equations of statistical equilibrium for the C stage.
(vi) Repeat step (ii) etc., calculating the u terms from the improved p values, and vice versa, until successive p values agree within specified limits, and the average traffic carried by both stages is consistent.

The table below shows a comparison made by Elldin[14] between the three foregoing methods of calculation and a traffic simulation test carried out by Wallstrom.[93] The trunking scheme is that of Fig. 34, with the following data:

$n = m = 4$ (i.e. no expansion or concentration)

$\phi = 0.1, a = 1$

$C = 2.3$ E (this is a nominal quantity, the value of which varies slightly according to the method of calculation)

Table 8

Method	Time congestion	Call congestion
Direct solution of equations of state	0·2217	0·1509
Jacobaeus's method	0·3488	0·2544
Iterative method, allowing for interstage dependence	0·2399	0·1734
Simulation (95% confidence limits)	0·2057—0·2443	0·1564—0·1676

The following iterative method for any number of switching stages is due to Bininda and Daisenberger.[15] Consider a link system of s stages in series. We are interested in the probability of not being able to connect a particular inlet of the first stage to any outlet of the last stage which is in the required route. Only those links which, if free and not blocked by traffic in other stages, could carry the call in question need be considered.

Let N_i = total number of links from stage i available to the call under consideration if free and unblocked.

For example, in Fig. 4b, for a call between two subscribers in group I, we have $N_1 = N_2 = N_3 = 3$, while in Fig. 5b $N_1 = 3, N_2 = 5, N_3 = 3$.

Let

A_i =   Traffic offered to the N_i links from stage i

$p_i(x)$ =   probability that x links out of N_i are busy

$\phi_i(y)$ =   prior probability that exactly y links from stage i are free and unblocked to the call in question by calls in *earlier* stages; the possibility of blocking by calls in later stages is left open

$\psi_i(y|x, z)$ =   conditional probability that exactly y of the $N_i - x$ free links from stage i are not blocked to the call in question by calls in earlier stages, given that z links from stage $i - 1$ are free and unblocked (not defined for $i = 1$).

Then

$$\phi_i(y) = \sum_{x=0}^{N_i} \sum_{z=0}^{N_{i-1}} \phi_{i-1}(z) p_i(x) \psi_i(y|x, z)$$

Clearly,

$$\phi_1(y) = p_1(N_1 - y)$$

Some combinations of x, y, and z may be impossible, depending on the trunking, in which case we put $\psi_i(y|x, z) = 0$. Assuming random allocation of links, the nonzero values of the ψ terms can be calculated combinatorially. As a simple example, consider a series–parallel network with $N_i = N$ at all stages (see Fig. 4 where $N = 3$). The number of ways in which the $N - x$ free links in stage i are arranged, so that y of them are aligned with free and unblocked links in stage $i - 1$, is obtained as follows.

First allocate $z - y$ calls in stage i to links which are connected to free and unblocked links in stage $i - 1$; these links can be chosen in $\begin{pmatrix} z \\ z - y \end{pmatrix}$ ways. Then allocate the remaining $x - (z - y)$ calls to links which are connected to busy or blocked links in stage $i - 1$; these links can be chosen in $\begin{pmatrix} N - z \\ x - z + y \end{pmatrix}$ ways.

Hence

$$\psi_i(y|x, z) = \frac{\binom{z}{y} \binom{N-z}{x-z+y}}{\binom{N}{x}}$$

Fig. 39b illustrates a possible arrangement for the case $i = 3$, $N = 9$, $x = 4$, $y = 2$, $z = 4$.

Initial values of the p_i terms are calculated from the assumed distributions for each stage; the ϕ_i terms can then be calculated stage by stage. The probability of blocking is $\phi_s(0)$. Improved values of the p_i terms can then be obtained from equations of statistical equilibrium, as explained in connection with the 2-stage scheme above, and the calculation repeated until the required degree of accuracy is obtained. In effect, the originally postulated distribution are thus modified iteratively until consistency between different stages is achieved. A further correction is necessary to allow for the direct dependence between the traffic in a switching stage and that part of it, in the next stage, which affects the route under consideration; in the 2-stage method described above, this was allowed for by means of the factors ϕ and θ. For this and other details of the method, reference should be made to the original paper.[15]

6.3 Effective availability

The concept of effective availability of a link system, which was developed independently by Bininda/Wendt and Kharkevich.[12,13] has been widely employed for approximate traffic calculation. It can be applied in different ways; the following exposition is based mainly on Lotze's c.i.r.b. (combined-inlet-and-route-blocking method).[86]

In the 2-stage link system of Figs. 33 and 34, the blocking can be divided into two parts, i.e.

(i) inlet blocking B_I, due to all the B devices which serve the calling line being busy;

(ii) route blocking B_R, due to all the C devices which are available to the calling line, via free B devices, being busy.

If $n \leqslant m$, and a is the average traffic carried per source, the inlet blocking is a^n in the sense of time congestion, i.e. all inlets or outlets of a switch in the first stage being occupied, and zero in the sense of call congestion, since calls cannot be lost in this state as there are no free sources, If $n > m$, the inlet blocking, in either sense, can be calculated from the Erlang or Engset formulas as appropriate, the B column being treated as a full-availability group, offered traffic by n sources.

The route blocking is determined as follows: let ϕ denote the proportion of traffic from the B column under consideration, which is directed to the route under consideration, and let b denote the average traffic carried per B device. On the average, $mb\phi$ devices in the B column are occupied by calls to the required route, and $m(1-b)$ are free. Hence, the average number of C devices in the required route which are available to the particular B column is

$$K = mb\phi + m(1-b)$$

In its simplest form, the method treats the route as equivalent to a grading of m trunks at availability K, the traffic offered to the equivalent 'grading' being the total traffic offered to the route. The route blocking can be determined from the m.P.J. or other appropriate traffic-capacity formula for gradings (Chapter 4). The same principle can be applied to any number of switching stages (Fig. 40a).

Let

s = number of stages

k_j = number of outlets from stage j, accessible to one inlet of that stage, and giving access to the required route

a_j = traffic carried per inlet of stage j (or outlet of stage $j-1$).

It is assumed that there is not more than one link between each pair of switches. Each inlet of stage 1 has access to k_1 inlets of stage 2, of which, on the average, $k_1(1-a_2)$ are free. Each of these has access to an average of $k_2(1-a_3)$ free inlets to stage 3, and so on. In a simple fanning-out scheme (Fig. 40b), this would mean that each inlet of stage 1 had access, on the average, to

$$k_1(1-a_2)\,k_2(1-a_3)\,...\,k_{s-1}(1-a_s)$$

separate free paths, each giving access to a separate set of k_s trunks in the required route. In the more typical pattern of Fig. 40a, however, some of these paths may overlap. Taking account of this, and of links in the route which are already occupied by calls from the inlet group under consideration, the average availability can be expressed as follows:

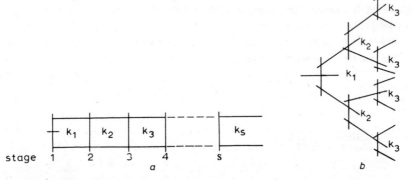

stage 1 2 3 4 s

a

b

Fig. 40 Effective availability

$$k = k_1 a_2 \phi + k_1 k_2 k_3 \dots k_s(1 - a_2)(1 - a_3) \dots (1 - a_s)$$

or the number of trunks in the route, whichever is the least. ϕ is defined as before. k_s is the number of links in the route which are accessible from one switch in the last stage; in other words, the number of columns in the route, which was 1 in the previous 2-stage example. The inlet and outlet blocking are determined as before; the former being based on the first stage only.

It is possible to regard any stage in a link system as equivalent to a grading, not only the final stage. If the method were exact, it would make no difference which stage was selected. Let N be the number of links, at some intermediate stage, which can form part of a connection between a particular pair of calling and called lines. The average number of these links which are accessible from the calling line via free paths can be calculated as before, and likewise the average number with access to the called line via free paths; let these quantities be denoted K_G and K_D respectively. The probability that a particular link is accessible from both ends is

$$\left(\frac{K_G}{N}\right)\left(\frac{K_D}{N}\right)$$

so the average number of accessible links out of N is $K_G K_D / N$. Allowing for links already occupied by calls between the particular inlet groups containing the calling and called lines, and denoting this traffic by y, the average availability becomes

$$y + \frac{K_G K_D}{N}$$

The effective availability method has been found to give good agreement with simulation tests on a 2-stage link system,[9] but it would be inadvisable to apply the method, or indeed, any other approximate method, to an untried trunking scheme without checking its accuracy, except for the purpose of very rough preliminary calculations. In principle, there is no prior reason for expecting that the effective availability of the equivalent grading should equal the average availability, and it may be possible to improve the accuracy of the method by an empirical modification of the availability. As we have seen, it is sometimes fairly easy to calculate the congestion by Jacobaeus's method assuming Bernoulli distribution at all stage. If B is the probability of congestion calculated in this way, and c is the average traffic per link in the route, the effective availability K is given by $c^K = B$. Applying this value of K to the equivalent grading, we can now recalculate the congestion using a suitable traffic-capacity formula, such as the m.P.J., which assumes a more realistic traffic-distribution in the route than the Bernoulli.

6.4 Geometric-group concept

This concept was suggested by Bridgford and Smith,[65,87] Consider a route of n trunks, accessible via a link system of any number of stages. Let $B(x)$ be the conditional probability that, if x trunks in the route are busy, access from a particular inlet to all free trunks is blocked by traffic in the link system. Clearly, $B(n) = 1$, and $B(n - 1)$ is usually fairly simple to calculate. Put $B(n - 1) = a$, and replace $B(n - 2)$, $B(n - 3)$ etc. by the geometric series a^2, a^3, etc. The probability of blocking becomes

$$\sum_{x=0}^{n} p(x)a^{n-x}$$

where $p(x)$ is the probability of x calls in the route. If we assume Erlang or Bernoulli distribution, this reduces to one of Jacobaeus's formulas. More accurately $p(x)$ can be determined from equations of statistical equilibrium, which give the following recurrence relationship (see Section 3.1); Poisson input is assumed in this example:

$$p(x + 1) = \frac{A(1 - a^{n-x})p(x)}{x + 1} \quad (A = \text{traffic offered})$$

which, together with the condition that the sum of the $p(x)$ is unity, enables all the state probabilities to be calculated.

The accuracy of the method depends on the trunking scheme. When applicable, it is probably intermediate in accuracy between Jacobaeus's method and recursive modification of postulated stage distributions (Section 6.2).

6.5 Link systems and overflow traffic

The variance of the overflow traffic from a link system can be calculated by treating it as a grading with the appropriate effective availability. Instead of taking the mean of the upper and lower limits as described in Section 4.9, however, Herzog states that the lower limit alone gives good results.[148] If 'peaky' nonPoisson traffic of known mean and variance, such as alternatively routed junction traffic (Section 11.3), is offered to a link system, the resulting congestion can be calculated by replacing the latter by a grading of the same effective availability, and applying the method described in Section 4.10. The effective availability may be assumed independent of the type of traffic offered, and calculated in the ordinary way. Alternatively, the geometric group concept can be adapted to calculations involving overflow traffic.[65]

6.6 Modes of connection in a link system

In applying the foregoing methods of calculation to traffic engineering, it is necessary to consider a number of modes of connection, including the following (Fig. 41). ·

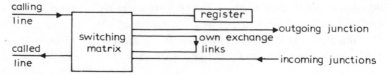

Fig. 41 Modes of connection in a link system

(i) *Incoming call*
It is assumed that an incoming or bothway junction has already been seized from the distant exchange, so that congestion on the junction route itself does not enter into the calculation, so far as the exchange under consideration is concerned. It is therefore only necessary to consider a connection from a particular inlet (the junction), to a particular outlet (the called line). This does not apply to systems in which the junction route, the switching network at the terminating exchange, and possibly that at the distant exchange, form part of the same conditonal-selection system; such a system should treated as a whole in congestion calculations.

(ii) *Connection to a register*
There is normally a common group of registers, accessible via one or more switching stages. Hence we require the probability that a particular inlet (the calling line) is unable to be connected to any one of a group of outlets (the registers), either because the latter are all busy or because there is no free path. Thus, in Fig. 34, the C column might represent the registers and the B column a switch group giving access thereto. The registers are usually worked on a delay basis, and could be treated in accordance with Chapter 7. Since, however, the register holding time is usually long compared with the tolerable dealy, they are often treated on a lost-call basis for convenience. Another possible arrangement is for calls to wait when all registers are busy, but to receive busy tone and be lost if there is a free register but no free path to it.

(iii) *Outgoing call*
If the dialled code is that of another exchange, connection is required to any free junction in that route. In some systems, the path already established to the register may affect the paths available for connection to a junction, and therefore the probability of congestion. Possible schemes include the following:

(*a*) If access to register and junction is obtained via completely separate networks, the register path does not affect the probability of connection to a junction.
(*b*) If the system compels the call to use the register access path as part of the final connection, the probability of congestion is calculated from the point at which the final connection diverges from the register access path.
(*c*) If there is a common network for register and junction access, but the call is obliged to set up a different path to the junction, the original path being released at the same time as the register, the number of available links in the relevant stages is reduced by one.

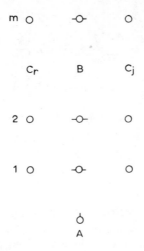

Fig. 42 Connection to register and junction

(*d*) If the call can use part of the register-access path to complete the connection, but is not compelled to do so, the distribution law can be modified in the light of the knowledge that at least one path is available up to a certain point.

A simple example is illustrated in Fig. 42. The B column gives access to two C-columns, the right-hand one representing junctions and the left-hand one registers. The call under consideration has just seized register no. 1, and we require the probability that it cannot now get access to a junction. B link no. 1 is already available for this purpose of calculation; in other words, we are concerned with the probable state of the system as it was just before the link was seized, neglecting any other changes which may have occurred during the short time between seizure of a register and application for a junction.

Let $p_B(x) =$ prior probability that x B devices are busy, $p_B(x| < m) =$ conditional probability that x B devices are busy, given that fewer than m are busy. Since at least one B device is free, the probability that x others are busy is $p(x| < m)$, which is given by the ratio of the prior probability that x devices are busy to the total prior probability that fewer than m are busy (Appendix 1.3). Thus

$$p_B(x| < m) = \frac{p_B(x)}{\sum\limits_{r=0}^{m-1} p_B(r)}$$

Now let $p_J(p) =$ probability that p junctions are busy.

Assuming random allocation of links, the probability that the $m - p$ B devices in the same row as the free junctions are all occupied is

$$H(m-p) = \sum_{x=m-p}^{m-1} p_B(x| < m) \frac{\binom{p}{x-m+p}}{\binom{m}{x}}$$

The probability of junction access being blocked is therefore

$$\sum_{p=0}^{m} H(m-p)\, p_J(p)$$

(iv) *Own-exchange call*
This type of call may be connected to any one of a group of internal links which is accessible from both the calling and the called line. Similar remarks about the effect of the register connection apply as in the case of outgoing calls. The effect of the 'doubling back' of own-exchange calls through the switching network can usually be neglected unless the calling and called lines both obtain access to the network via a small group of bothway links. The proportion of such calls is usually very small, and a relatively inferior grade of service may be tolerable for that reason, but they obviously cannot be ignored. A reasonable approximation is to calculate congestion as if there were different links groups at the calling and called end, the latter having one fewer link than the former.

6.7 Call packing

In conditional selection systems, as in gradings, blocking depends to some extent on the order in which links are allocated. Thus, in Fig. 3, it is more efficient to allocate the B switches sequentially, starting from a fixed point, then at random. This ensures that connected pairs of A-B and B-C links are left free as long as possible. It is possible to devise more complicated allocation rules to place each call where it is least likely to cause blocking, but this is unlikely to produce very much improvement. The increase in traffic capacity as compared with random allocation lies between 1% and 10%, depending on the trunking scheme. These results have been established by simulation; the effect of call packing is difficult to calculate theoretically.[39]

6.8 Common control

A group of switches with their common controls (Section 1.5) forms a link system, and the usual methods of calculation are applicable. For example, the arrangement shown in Fig. 8 can be represented in conventional link diagram form as in Fig. 43,

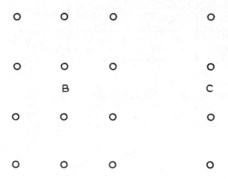

Fig. 43 Simplified form of Fig. 8

where the B stage represents the speech paths and the C stage represents the controls. This does not mean that the speech paths are necessarily seized before the controls, which depends on the mode of operation of the system, and is irrelevant to the method of calculation; the stages in Fig. 43 could equally well be reversed.

If the traffic in each *B* column is B erlang, the total C stage traffic

$$C = 3Br \text{ erlang}$$

where

$$r = \frac{\text{average control-holding time per call}}{\text{average switch-holding time per call}}$$

The switch-holding time includes the time for which the switch is associated with the control. The appropriate formula of Jacobaeus can be applied, depending on the distribution postulated. If the number of sources is large, Erlang distribution is applicable to the switches and, strictly speaking, to the controls also. If, however, the control traffic is so small that the chance of all controls being occupied simultaneously is negligible, Bernoulli distribution may be sufficiently accurate, which, as we have seen, often leads to simpler formulas.

An alternative formula, due to G.S. Berkeley[50] treats the periods when a switch is unavailable, due to its control being associated with another switch, as equivalent to artificial traffic on the switch, the amount of which is calculated as follows:

Let

k = average number of switches served by one control

g = average traffic carried by one switch, including times when the switch is associated with its control

h = average traffic carried by a control due to one switch.

Consider the states of a particular switch during a long period T. It is occupied for a total time gT, and is associated with its control for time hT. The control is associated with other switches for time $(k-1)hT$, and this state may be assumed

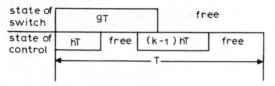

Fig. 44 Effect of common control on switch availability

randomly distributed over those times, totalling $T - hT$, when the switch under consideration is not associated with its control, including $T - gT$ when it is free. Hence, the time during which the switch is free and its control is engaged is

$$(k - 1)hT \frac{T - gT}{T - hT}$$

Dividing this by T we obtain the artificial traffic

$$f = \frac{(k - 1)h(1 - g)}{1 - h}$$

The relationship of the different states is shown diagrammatically in Fig. 44; for convenience, the time occupied by each state is shown as if it were continuous.

If the switch is one of a group of s switches carrying y erlang, and producing b erlang control traffic, and the control is one of a group of c switches carrying a erlang, in all, we have

$$y = sg, a = kch, b = sh$$

The effective traffic carried by the group of s switches is

$$y + sf = y + \frac{sa(k - 1)(s - y)}{kc(s - b)}$$

At normal grades of service, the difference between offered and carried traffic can often be neglected. Congestion is calculated by the appropriate formula (e.g. Erlang's lost-call formula if the input is Poissonian and the s switches constitute a full-availability group), using the effective traffic instead of the real traffic y; this usually overestimates the congestion, if anything, so the error is on the safe side. The method is simple to use, as it does not take account of the particular pattern of connection between switches and controls. Given y, the value of $y + sf$ is calculated for trial values of s until the number of trunks required ot carry $y + sf$ erlang, as read from the appropriate traffic-capacity table, agrees with s. If, as occasionally happens, $s - 1$ is too low and s is too high, then s trunks should be provided.

In schemes such as that of Fig. 9, where any switch can be coupled to any control, it is usually accurate enough to treat the controls as a separate full-availability group, and to apply Erlang's loss or delay formula as appropriate, using the control traffic as a parameter.

It will be noticed that, in Fig. 8, the controls are arranged so that each group of switches is served by all controls; this is more efficient than dividing the controls into separate groups, since that would increase the chance of all controls which serve a group of switches being busy simultaneously, on the same principle that a small trunk group is less efficient than a large one (Chapter 3).

6.9 Crosspoint minimisation

The crosspoints usually account for a high proportion of the cost of a link system, so that the design of networks requiring minimum numbers of crosspoints is an important aspect of telecommunication-systems engineering. This is not, however, an absolute optimisation criterion, since it may conflict with other important considerations, such as standardisation of switch sizes. Moreover, it does not directly take account of costs which are associated with control functions and link circuits rather than individual crosspoints.

The following approximate method of minimisation is due to Lotze.[88] It is applicable to various types of link system, with or without grading between stages. It is known as the c.p.e. (crosspoints-per-erlang) method. The average availability of a route to which access is obtained via a link system of s stages includes the term

$$T = k_1 k_2 \dots k_s (1 - a_2)(1 - a_3) \dots (1 - a_s)$$

which Lotze calls the 'transparency'. In the present context, k_s is the number of outlets to which each inlet of the last stage has access, irrespective of whether or not they are all in the same route, as this does not affect the result. As in Section 6.3, it is assumed that there is not more than one link between each pair of switches.

Each inlet in stage j requires k_j crosspoints and carries a_j erlang, so that the average number of crosspoints per erlang for that stage is k_j/a_j. The quantity

$$\text{c.p.e.} = \frac{k_1}{a_1} + \frac{k_2}{a_2} + \dots + \frac{k_s}{a_s}$$

is a measure of the rate of provision of crosspoints in relation to traffic flow for the system, while T is a measure of the effective availability and therefore indirectly, of the grade of service. The object is to select switch sizes and number of stages so as to minimise c.p.e. while keeping T constant.

Putting $Q = T/a_1$ we have

$$\text{c.p.e.} = \frac{Q}{k_2 k_3 \dots k_s (1 - a_2)(1 - a_3) \dots (1 - a_s)} + \frac{k_2}{a_2} + \frac{k_3}{a_3} + \dots + \frac{k_s}{a_s}$$

a_1, being the source inlet loading, is assumed to be given.

Keeping T, s and a_1 constant, a necessary condition for a minimum value of c.p.e. is

$$\frac{\partial \text{c.p.e.}}{\partial k_2} = \frac{\partial \text{c.p.e.}}{\partial k_3} = \ldots \frac{\partial \text{c.p.e.}}{\partial k_s} = 0$$

$$\frac{\partial \text{c.p.e.}}{\partial a_2} = \frac{\partial \text{c.p.e.}}{\partial a_3} = \ldots \frac{\partial \text{c.p.e.}}{\partial a_s} = 0$$

Now

$$\frac{\partial \text{c.p.e.}}{\partial k_2} = -(Q/k_2 R) + (1/a_2)$$

where

$$R = k_2 k_3 \ldots k_s (1 - a_2)(1 - a_3) \ldots (1 - a_s)$$

Hence, for a minimum value of c.p.e., we have

$$k_2/a_2 = k_3/a_3 = \ldots = k_s/a_s = Q/R$$

Also

$$\frac{\partial \text{c.p.e.}}{\partial a_2} = \{Q/R(1 - a_2)\} - (Q/Ra_2) = 0$$

Thus

$$a_2 = (1 - a_2) = 0 \cdot 5 = a_3 = \ldots = a_s$$

therefore

$$k_2 = k_3 = \ldots = k_s \equiv k = 0 \cdot 5 Q/R = (2^{s-2} Q/k^{s-1})$$

thus

$$k = 2(Q/4)^{1/s}$$

Moreover, if i_1 is the number of inlets per switch in stage 1, we have

$$i_1 a_1 = k_1 a_2 = 0 \cdot 5 k_1 = \frac{0 \cdot 5 T}{(0 \cdot 5 k)^{s-1}}$$

thus

$$i_1 = \frac{0 \cdot 5 T/a_1}{(0 \cdot 5 k)^{s-1}} = \frac{0 \cdot 5 Q}{(0 \cdot 5 k)^{s-1}} = \frac{2(k/2)^s}{(k/2)^{s-1}} = k$$

and

$$k_1 = 2ka_1$$

It follows that the minimum value of c.p.e. is given by

$$\text{c.p.e.}_{min} = 4s(Q/4)^{1/s}$$

The optimum number of stages s_{opt}, is given by

$$\frac{d\text{c.p.e.}_{min}}{ds} = 0$$

whence

$$s_{opt} = \log_e(Q/4)$$

To apply the method, it is necessary to specify the minimum value of T. This can be determined from the route with the largest effective availability. The latter cannot be accurately determined in advance, because it depends on parameters which have still to be optimised. Approximate estimates can, however, be made using trial values. Subtracting the inlet blocking from the required total blocking gives the outlet blocking, and the corresponding average availability can then be found from the appropriate congestion formula, knowing the traffic on the route. Subtracting $k_1 a_2 \phi$ gives the transparency for this route; the total transparency T is obtained by multiplying this by k_s/k_s' where

k_s' = number of s-stage outlets in the route in question, accessible to each s-stage inlet (k_s' was denoted k_s in Section 6.3)

k_s = total number of s-stage outlets accessible to each s-stage inlet (as before).

The optimum values of the k-terms and s can now be calculated. They will probably require adjustment to integral values. If this process results in more than one alternative arrangement with the same total number of crosspoints, that with the highest value of T is optimum. As the method is approximate, the grade of service on each route should be checked. If it is desired to ensure negligible internal blocking in the link system, the value of T must be at least equal to the total numbers of trunks from the last stage, and it may be advisable to make it, say, 1·5 times that number.

Chapter 7

Waiting-call systems

At very low probabilities of congestion, provision of waiting facilities does not have much effect on traffic capacity, and, for convenience, is often ignored, calculations being made on a lost-call basis. There are circumstances, however, where relatively high congestion can be tolerated because calls are able to wait for service, and calculations should then be on a delay basis. This applies particularly to common-control equipment, where the holding time is relatively short, so that a high proportion of delayed calls may be tolerable without producing excessive waiting times. The grade of service in a waiting-call system is usually expressed in terms of the probability of exceeding a specified maximum delay. The average delay, which is generally easier to calculate, is also a useful measure for comparing the performance of different systems. It should, however, be used with discretion, since a low average delay may be compatible with an unacceptable proportion of very long delays.

In dealing with Erlang's lost-call formula a negative exponential distribution of holding times was assumed for mathematical convenience, although the formula can be proved to hold for any distribution (Section 3.1). This is intuitively reasonable, since, provided the call origins occur at random, their end-points are also random. In the case of a waiting-call system, however, this is not generally the case. Consider a single-trunk system, with all calls having the same holding time. If there is more than one call in the queue, they will receive service, and will therefore terminate, at regular intervals of one holding time. Queueing therefore tends to smooth out the end-points, except in the special case of the negative exponential distribution, where the future duration is independent of its past duration; this case is the simplest to analyse, and is the only one which will be dealt with in detail in this monograph.

7.1 Probability of delay

The case of a full-availability group with Poisson input can be analysed in a similar way to the lost-call system. Assumptions a, b and d of Section 3.1 are still applicable, but c is replaced by the following:

Calls originating when all trunks are busy wait for service as long as necessary, and are connected immediately when a trunk becomes free.

If N is the number of trunks and A is the traffic offered, it is a necessary condition of statistical equilibrium that $A < N$; otherwise, when all trunks are occupied, the average arrival rate is greater than or equal to the average termination rate, so that the queue builds up indefinitely. If $A < N$, however, all calls are eventually carried, so there is no difference between offered and carried traffic flow. If the system contains x calls, and $x \leqslant N$, so that x trunks are occupied, the probability of a call terminating during a short interval dt is, as before, $x dt$. If, however, $x > N$, so that $x - N$ calls are waiting, the probability is $N dt$. The probability of a call arrival is $A dt$ irrespective of the state of the system. The equations of statistical equilibrium are therefore

$$(A + x)\, p(x) \;=\; Ap(x-1) + (x+1)p(x+1) \quad \{0 \leqslant x < N, p(-1) = 0\}$$

$$(A + x)\, p(x) \;=\; Ap(x-1) + Np(x+1) \qquad (x \geqslant N)$$

where $p(x)$ is the probability of x calls being in the system simultaneously, including waiting calls.

Hence
$$p(1) \;=\; Ap(0)$$

$$p(2) \;=\; \frac{A^2}{2!}\, p(0)$$

$$p(3) \;=\; \frac{A^3}{3!} p(0)$$

etc. to

$$p(N) \;=\; \frac{A^N}{N!}\, p(0)$$

$$p(N+1) \;=\; \left(\frac{A}{N}\right)\frac{A^N}{N!}\, p(0)$$

$$p(N+2) \;=\; \left(\frac{A}{N}\right)^{\!2} \frac{A^N}{N!}\, p(0)$$

etc.

Moreover
$$\sum_{x=0}^{\infty} p(x) \;=\; 1$$

Thus
$$p(0) \;=\; 1 \Big/ \left\{ \sum_{r=0}^{N-1} \frac{A^r}{r!} + \frac{A^N}{N!} \sum_{r=0}^{\infty} \left(\frac{A}{N}\right)^{\!r} \right\}$$

$$=\; \left[1 \Big/ \left\{ \sum_{r=0}^{N-1} \frac{A^r}{r!} + \frac{A^N}{N!}\left(\frac{N}{N-A}\right) \right\} \right] (A < N)$$

The probability of congestion is usually denoted

$$E_{2,N}(A) = \sum_{x=N}^{\infty} p(x) = \left\{ \frac{A^N}{N!} \left(\frac{N}{N-A} \right) \right\} \Big/ \left\{ \sum_{r=0}^{N-1} \frac{A^r}{r!} + \frac{A^N}{N!} \left(\frac{N}{N-A} \right) \right\}$$

This is Erlang's delayed-call formula.

At low congestion, the last term of the denominator is relatively small and can be approximated by $A^N/N!$, so that

$$E_{2,N}(A) = E_{1,N}(A) \frac{N}{N-A} \quad \text{approximately.}$$

where $E_{1,N}(A)$ is Erlang's loss probability for the same values of A and N.

An accurate relationship between the loss and delay formulas is

$$E_{2,N}(A) = \frac{N E_{1,N}(A)}{N - A + A E_{1,N}(A)}$$

As A approaches N from below, so that the occupancy approaches one, the denominator of $E_{2,n}(A)$ is dominated by the last term, so that the probability of congestion approaches one, as would be expected.

In the case of a single-trunk group, the traffic carried is, by definition, equal to the probability that the trunk is occupied (Section 2.1). The formula agrees with this definition, since

$$E_{2,1}(A) = \frac{A\left(\dfrac{1}{1-A} \right)}{1 + A\left(\dfrac{1}{1-A} \right)} = A$$

The probability that all trunks are busy and there are j waiting calls is

$$p(N+j) = \frac{\dfrac{A^N}{N!} \left(\dfrac{A}{N} \right)^j}{\displaystyle\sum_{r=0}^{N-1} \frac{A^r}{r!} + \frac{A^N}{N!} \left(\frac{N}{N-A} \right)}$$

$$= E_{2,N}(A) \left(1 - \frac{A}{N} \right) \left(\frac{A}{N} \right)^j$$

The probability of all trunks being busy with j *or more* waiting calls is

$$\sum_{r=j}^{\infty} E_{2,N}(A) \left(\frac{A}{N} \right)^r \left(1 - \frac{A}{N} \right) = E_{2,N}(A) \left(\frac{A}{N} \right)^j$$

7.2 Average waiting time

The average number of waiting calls is

$$
\begin{aligned}
\bar{q} &= \sum_{x=N+1}^{\infty} (x-N)\,p(x) \\
&= \frac{A^N}{N!}\,p(0) \sum_{x=N+1}^{\infty} (x-N)\left(\frac{A}{N}\right)^{x-n}. \\
&= \frac{A^N}{N!}\left(\frac{A}{N}\right)\left(1-\frac{A}{N}\right)^{-2} p(0) \\
&= E_{2,\,N}(A)\left(\frac{A}{N-A}\right)
\end{aligned}
$$

Let $\bar{w}$ denote the waiting time of a call, averaged over all calls, including those which are not delayed. As usual we take the average holding time as the unit of time. During a long period Z, the average number of calls carried is AZ, and their total waiting time is $AZ\bar{w}$. This can also be expressed as $\bar{q}Z$, so that

$$
\bar{w} = \frac{\bar{q}}{A} = \frac{E_{2,\,N}(A)}{N-A}
$$

The delay, averaged over delayed calls only, is

$$
\bar{w} \text{ probability of delay } = 1/(N-A)
$$

The average duration of the congestion state, calculated as in Section 3.8, is also $1/(N-A)$, despite the fact that waiting times are shorter than the congestion states during which they occur. The explanation of this apparent paradox is that more calls are delayed during long periods than during short periods of congestion (see Section 2.5).

In the present context, waiting time means the time between the arrival of a call and seizure of the device under consideration. Whether this represents the delay experienced by the caller depends on the function of the device. If it carries speech or data, or generates an audible signal such as dialling tone, service may be considered to commence at the moment of connection. On the other hand, if it is a control device used in setting up the connection, its holding time may be an indistinguishable part of the experienced delay.

7.3 Distribution of waiting times

So far, no assumption has been made as to the order in which waiting calls are served, the *queue discipline*. The only discipline to be analysed in any detail in this monograph is 'first come first served', (f.c.f.s.) which is also known as 'first in first out' (f.i.f.o.) and 'order of arrival service'.

If all trunks are occupied, the probability that no termination occurs within a given time t, expressed as a multiple of the average holding time, is

$$(e^{-t})^N = e^{-Nt}$$

If there is at least one waiting call at the time of the first termination, so that the trunk is immediately reoccupied, the interval until the next termination has the same distribution, since it is independent of the ages of the calls. Thus, so long as all trunks are occupied, the intervals between terminations have negative exponential distribution with mean value $1/N$, so that the number of terminations in a given time t has Poisson distribution with mean value Nt (Section 2.5).

If a call arrives when N trunks are busy and j other calls are waiting, it will suffer a delay greater than t if and only if j or fewer calls terminate within a time t. These terminations may include calls which are waiting when the new call arrives as well as calls which are already being served. The probability of this is

$$\sum_{r=0}^{j} \cdot \frac{(Nt)^r}{r!} e^{-Nt}$$

The total probability of the delay exceeding t for all values of j is therefore

$$F(t) = \sum_{j=0}^{\infty} p(N+j) \sum_{r=0}^{j} \frac{(Nt)^r}{r!} e^{-Nt}$$

$$= E_{2,N}(A) \left(\frac{N-A}{N}\right) e^{-Nt} \left\{ 1 + \left(\frac{A}{N}\right) \left[1 + \frac{(Nt)^1}{1!}\right] + \left(\frac{A}{N}\right)^2 \left[1 + \frac{(Nt)^1}{1!} + \frac{(Nt)^2}{2!}\right] + \ldots \right\}$$

$$= E_{2,N}(A) e^{-Nt} \left\{ 1 + \frac{(Nt)^1}{1!} \left(\frac{A}{N}\right) + \frac{(Nt)^2}{2!} \left(\frac{A}{N}\right)^2 + \ldots \right\}$$

$$= E_{2,N}(A) e^{-(N-A)t}$$

This is the unconditional probability of delay greater than t; the conditional probability of delay, given that a call *is* delayed, is

$$\frac{F(t)}{F(0)} = \frac{F(t)}{E_{2,N}(A)} = e^{-(N-A)t} = e^{-N(1-a)t}$$

where $a = A/N$, the average trunk occupancy.

In other words, waiting times, like holding times, have negative exponential distribution; the mean value is

$$\text{average holding time}/N(1-a)$$

With queue disciplines other than f.i.f.o. however, the waiting time distribution is not exponential.

Fig. 45 shows a typical set of delay probability curves for five trunks. The unconditional probability of delay exceeding a specified value t is plotted against t for a range of trunk occupancies a. The intercept on the vertical axis $F(0)$ is $E_{2,N}(A)$.

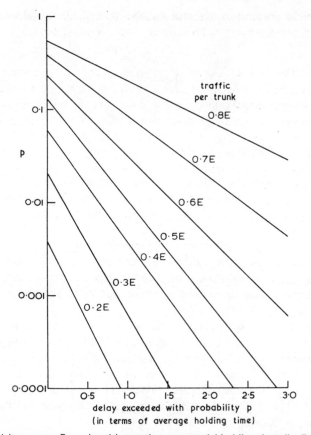

p

traffic
per trunk

0·8E

0·7E

0·6E

0·5E

0·4E

0·3E

0·2E

delay exceeded with probability p
(in terms of average holding time)

Fig. 45 Typical delay curves: 5 trunks with negative exponential holding-time distribution

In most applications, we are primarily interested in the probability of exceeding a specified delay in seizing a trunk If necessary, however. the distribution of the total time from arrival until completion of a call can be obtained by convoluting the distributions of the waiting and holding times as follows:

Let $f(t)$ = probability-density function (p.d.f.) of the waiting time;
$G(t)$ = cumulative distribution function of the holding time.
B = probability of congestion

The probability that the waiting time lies between t and $t + dt$ is $f(t)dt$, and the probability that the subsequent holding time exceeds $T - t$, where T is the specified tolerable delay until completion, is

$$1 - G(T - t)$$

The probability that the call is not completed within time T is therefore

$$(1 - B)\{1 - G(T)\} + \int_{t=0}^{T} f(t)\{1 - G(T - t)\}dt + \int_{t=T}^{\infty} f(t)dt$$

In the case of negative exponential distribution of holding and waiting times, we have

$$f(t) = E_{2,N}(A)e^{-(N-A)t/h}(N-A)/h$$
$$1 - G(T-t) = e^{-(T-t)/h}$$

where h = average holding time.

7.4 Constant holding time

The case when all calls have the same holding time was analysed by Pollaczek[89] and Crommelin,[21] independently. It is of practical interest for two reasons. First, many delay problems are concerned with common-control devices, the holding time of which may be independent of subscribers' behaviour and approximately constant. Secondly, the past duration of a call completely determines its future duration, so this case is, in a sense, the antithesis of the memory-lacking negative exponential distribution. If the distribution is unknown or has not been theoretically analysed, these extremes can often be used to obtain optimistic and pessimistic estimates, respectively. .

The probability of congestion is generally the same, to a close approximation, as for the negative exponential case; agreement is exact for a single-trunk system, since the probability is equal to the traffic carried by definition. Fig. 46 shows a comparison between typical delay curves for these two holding-time distributions. It will be noticed that the 'constant' curves exhibit a marked cusp at each integral number of holding times. First-in-first-out queue discipline and Poisson input is assumed in both cases.

7.5 Random service

The delay distribution, but not the average delay, depends on the order in which calls are served. Queuing in strict order of arrival may require expensive equipment, in which case a simpler method may be preferable, such as scanning in cyclic order or 'gating', in which calls are admitted in batches and then served in random order.[91] The case of purely random service is of practical interest, because many switching devices approximate thereto, and also as a limiting case. Comparison between the delay distributions with first-in-first-out service and random service gives a rough idea of the possible benefit of queuing, which may be sufficient for an engineering decision.

The case of negative exponential distribution has been analysed by Riordan,[22, 92] and working curves have been computed by Wilkinson.[23] Burke has analysed and computed working curves for the case of constant holding time.[25] A typical curve for the latter case is shown dotted in Fig. 46. Serving calls out of turn obviously

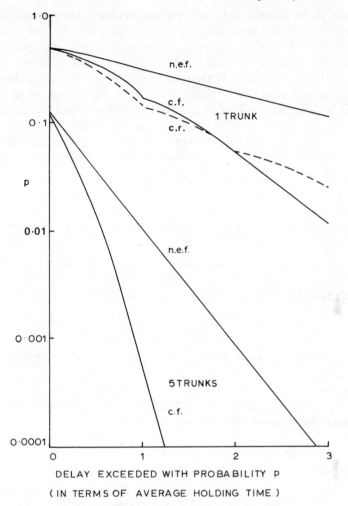

Fig. 46 Delay curves: effect of holding-time distribution and order of service
n.e.f. = negative exponential holding times, first-in-first-out
c.f. = constant holding times, first-in-first-out
c.r. = constant holding times, random service
Average traffic per trunk = 0·5 E in all cases

increases the chance of both very long and very short delays; this is reflected in the intersection of the random and f.i.f.o. curves.

Table 9 below shows the effect of different holding-time distributions and queue disciplines on permissible loading of trunks. It shows the traffic capacity of a single device with 1 call in 100 delayed more than 2·5 s with different average holding times. If the average is very short, the average queue length can be very large without the tolerable delay being exceeded, so that a high traffic loading can be achieved. If the tolerable delay is small compared with the average holding time, a call which finds all trunks occupied will probably have to wait longer than the tolerable limit before an existing call terminates; hence, the probability of an unacceptable delay is not much less than the probability of finding all the trunks busy, which is independent of the queue discipline and practically independent of the holding-time distribution. If the tolerable delay is very long compared with the average holding time, an occupancy of nearly 100% can be achieved, so that queue discipline and holding-time distribution have little effect. Thus, the advantage of queuing is most marked in an intermediate region, the precise location of which depends on the holding-time distribution. These conclusions are equally applicable to waiting systems with more than one device.

Table 9

Holding-time distribution	Order of service	Average holding time		
		0·1 s	1 s	10 s
Exponential	1st in, 1st out	0·82 E	0·10 E	0·014 E
	random	0·73 E	0·10 E	0·014 E
Constant	1st in, 1st out	0·91 E	0·40 E	0·014 E
	random	0·86 E	0·33 E	0·014 E

7.6 Single trunk: general holding-time distribution

In the special case of a single-trunk system with Poisson input and first-in-first-out service, a simple expression for average delay can be derived as follows:

Let $b(t)$ = probability-density function of the holding-time distribution
$\bar{t}$ = average holding time
A = traffic in erlangs.

Consider a very long period Z. The average number of calls during Z is $AZ/\bar{t}$, since A is, by definition, the average number of calls during $\bar{t}$. The average number of calls during Z with holding times in a specified short range, $t - t + dt$, is therefore

$$\frac{AZ}{\bar{t}} b(t) dt$$

The average time occupied by calls of this particular holding time is

$$\frac{tAZb(t) dt}{\bar{t}}$$

while the average time occupied by all calls is

$$\frac{AZ}{\bar{t}}\,\bar{t} \;=\; AZ$$

If a call finds the trunk occupied, then since it is equally likely to arrive at any time during the service of the existing call, the average interval before the latter ends, if its total holding time is between t and $t + dt$, is $t/2$. But the probability that the new call encounters an existing call with this particular holding time is the ratio of the time occupied by these calls to the time occupied by all calls, which is $tb(t)dt/\bar{t}$. The interval before the next call termination, averaged over all holding times, is therefore

$$\int_{t=0}^{\infty} \left(\frac{t}{2}\right)\left(\frac{tb(t)dt}{\bar{t}}\right) \;=\; \frac{1}{2\bar{t}} \int_{t=0}^{\infty} t^2 b(t)\, dt$$

$$=\; \frac{1}{2\bar{t}}\,[s^2 + \bar{t}^{\,2}]$$

where s is the standard deviation of the holding time t.

The waiting time of a call comprises the interval until the end of the existing call plus the holding times of the previous waiting calls, if any. As before we denote the number of waiting calls, averaged over occupied and unoccupied times, by $\bar{q}$, so the average number of waiting calls during occupied times is $\bar{q}/A$. Hence, the average waiting time of *delayed* calls is

$$\frac{1}{2\bar{t}}\,[s^2 + \bar{t}^{\,2}] + \frac{\bar{q}}{A}\,\bar{t}$$

Moreover, the waiting time averaged over all calls is

$$\bar{w} \;=\; \bar{q}\,\bar{t}/A$$

(see Section 7.2, where $\bar{t}$ was taken as 1 for convenience). Substituting for $\bar{q}$, the above expression for the average waiting time of delayed calls becomes

$$\frac{1}{2\bar{t}}\,[s^2 + \bar{t}^{\,2}] + \bar{w}$$

The waiting time averaged over all calls is obtained by multiplying this expression by the probability of delay A. Hence, we have

$$\bar{w} \;=\; \frac{A}{2\bar{t}}\,[s^2 + \bar{t}^{\,2}] + A\bar{w}$$

Thus
$$\bar{w} \;=\; \frac{A(1 + C^2)\,\bar{t}}{2(1 - A)}$$

where $C = s/\bar{t}$ (the *coefficient of variation* of the holding time). This formula is due to Pollaczek and Khinchine, independently. It follows that the average delay with constant holding time $(C = 0)$ is half that with negative exponential holding time $(C = 1)$.

In general, the waiting-time distribution cannot be expressed explicitly in terms of the holding-time distribution function; when required, delay distributions for empirical holding-time distributions are usually derived by simulation or approximate methods. An arbitrary holding-time distribution can often be approximated by means of Erlang's 'method of stages', in which a call is regarded as made up of a number of fictitious stages, each having negative exponential distribution.[56] Thus each stage, though not the call as a whole, exhibits the 'lack of memory', property which considerably simplifies theoretical analysis. This distribution is known as Erlangian or gamma. A somewhat similar device is the so-called method of parallel stages, in which the calls are regarded as a mixture of two or more populations, each having negative exponential distribution with different means; the resulting distribution is known as hyperexponential.

7.7 Finite capacity queues

So far, it has been assumed that there is no limit to the number of calls which the queue can accommodate. In practice, this is not the case, and busy tone is returned when the queue is full. If K is the queue capacity, N is the number of trunks and A is the traffic offered, the equilibrium equations for Poisson input and negative exponential holding-time distribution are

$$(x + 1)p(x + 1) = Ap(x) \quad (0 \leqslant x < N)$$

$$Np(x + 1) = Ap(x) \quad (N \leqslant x < N + K)$$

$$p(x) = 0 \quad (x > N + K)$$

The cases $K = 0$ and $K = \infty$ correspond to pure loss and delay systems, respectively. The following expressions are obtained for the probability of finding x calls in the system.

$$p(x) = \frac{A^x}{x!} p(0) \quad (0 \leqslant x \leqslant N)$$

$$p(x) = \left(\frac{A}{N}\right)^{x-N} \frac{A^N}{N!} p(0) \quad (N < x \leqslant N + K)$$

$$p(0) = \cfrac{1}{\displaystyle\sum_{r=0}^{N-1} \frac{A^r}{r!} + \frac{A^N}{N!} \left(\frac{N}{N-A}\right) \left[1 - \left(\frac{A}{N}\right)^{K+1}\right]}$$

Unlike the pure delay system, the combined delay and loss system attains equilibrium even if $A > N$. Expressions for average delay etc. can be derived in the same way as for the unlimited queue. The probability of loss, of course, is $p(N + K)$. In the case of a single-trunk system, this is

$$p(1 + K) = \frac{A^{1+K}(1-A)}{1 - A^{2+K}}$$

The carried traffic is

$$A[1 - p(1 + K)] = \frac{A(1 - A^{1+K})}{1 - A^{2+K}}$$

A common practical problem is to determine the queue capacity required to ensure that the probability of loss does not exceed a specified value. If the tolerable loss is small, as is usually the case, it may be sufficiently accurate to calculate $p(N + K)$ as if the queue were unlimited. This slightly underestimates the loss. An upper bound might be obtained by taking the loss as being

$$p(N + k) + p(N + k + 1) + p(N + k + 2) + \ldots = E_{2,N}(A)(A/N)^K$$

the queue being regarded as unlimited (Section 7.1).

7.8 Queues with priorities

In some systems, particularly in message switching, demands can be handled in order of priority, irrespective of their order of arrival. Priority may be pre-emptive or nonpre-emptive, according to whether a call can or cannot interrupt another of lower priority. Only a few simple cases will be considered here.

Consider a single trunk with any number of nonpre-emptive priority classes $1, 2, \ldots k$; class $k - 1$ takes precedence over class k. Poisson input is assumed.

Let A = total traffic
A_k = traffic in class k
h_k = average holding time for class k
h = holding time averaged over all classes
C = coefficient of variation of h (i.e. ratio of standard deviation to mean).

It can be shown[94] that the waiting time averaged over all calls in class k is

$$\overline{w}_k = \frac{A(1 + C^2)h}{(1 - \sum_{r=1}^{k} A_r)(1 - \sum_{r=1}^{k-1} A_r)} \cdot \frac{1}{2}$$

The average number of waiting calls of class k is $A \overline{w}_k / h_k$.

Cohen has analysed the case of N trunks with two nonpre-emptive priority classes, Poisson input and negative exponential holding-time distribution.[95] The probability of a call in either class being delayed is, of course, $E_{2,N}(A_1 + A_2)$. Assuming that both classes have the same average holding time, the average number of waiting calls in classes 1 and 2 are $\dfrac{A_1 E_{2,N}(A_1 + A_2)}{N - A_1}$ and $\dfrac{NA_2 E_{2,N}(A_1 + A_2)}{(N - A_1)(N - A_1 - A_2)}$, respectively. The waiting times averaged over delayed calls in classes 1 and 2 are $\dfrac{1}{N - A_1}$ and $\dfrac{N}{(N - A_1)(N - A_1 - A_2)}$, respectively. Thus, the average waiting time for class 1 is the same as if it were the only class of call. This is not generally true for nonpre-emptive priority systems with other holding-time distributions. In the case of pre-emptive priority, however, the highest class behaves in every respect as if it were the only class, for any holding-time distribution. Pre-emptive priority is applicable to pure loss as well as to delay systems.

Wagner[96] has dealt with nonpre-emptive priority systems with limited queue capacity. A special case of nonpre-emptive priority arises when calls encountering congestion are lost or delayed according to their source. For example, a group of trunks may be shared by manual operators who can continuously observe the state of traffic, and connect a call when a trunk is available, and subscribers dialling direct, who merely receive busy tone in the event of congestion. To understand the behaviour of this system, it is instructive to consider two limiting cases, as follows:

Let classes 1 and 2 denote calls with and without delay facilities, respectively. Suppose class 2 consists of a very large number of very short calls. If all trunks are occupied. some or all of them by class 2 calls, a new call in class 1 receives practically immediate service. Hence, the presence of class 2 calls has negligible effect on class 1 calls. It follows that the system behaves, with respect to class 1 calls, as if they were the only traffic; class 2 calls can occupy trunks only when not required by class 1 calls.

Suppose, on the other hand, that class 2 consists of a very small number of very long calls. If x trunks are occupied by class 2 calls, this state will persist for a long time, during which, from the point of view of class 1, the system behaves like a pure delay system of $N - x$ trunks. If $N - x$ is less than the class 1 traffic in erlangs, there will be no equilibrium and the queue will grow indefinitely.

Pratt[97] has determined the probability of loss for class 2 calls, assuming Poisson input and negative exponential holding-time distribution. The computation is facilitated by the fact that, in practice, the upper and lower limits, corresponding to these two extreme cases, are fairly close together.

If both classes have the same average holding time, the following expressions are obtained:[95]

Probability of blocking (loss or delay)

$$= \frac{NE_{1,N}(A_1 + A_2)}{N - A_1 + A_1 E_{1,N}(A_1 + A_2)}$$

Delay averaged over all calls with delay facilities, and expressed in terms of average holding time, is

$$\frac{(\text{probability of blocking})}{N - A_1}$$

The probability of delay exceeding t average holding times, for a call with delay facilities, is

$$E_{1,N}(A_1 + A_2)e^{-(N-A_1)t}$$

Gosztony[99] has dealt with a similar loss-delay system with a limited number of traffic sources and limited queue capacity.

7.9 Waiting systems in tandem

Fig. 47 shows two successive switching stages, each with its own queue. Suppose, in the first instance, that these form part of a message switching network of the store-and-forward type. If there is no free outlet from stage 1 in the required route, the message enters queue Q_1. When an outlet is available, there may be no free outlet from stage 2; in that case, the message enters queue Q_2, and the link between stages 1 and 2 is released. With Poisson input and negative exponential holding time, it can be proved that the input to stage 2 is also Poissonian, irrespective of congestion in stage 1.[100] Thus, the two queues can be treated as independent in calculating delays. The total probability of a call being delayed, and the average value of the total delay, are the sums of the probabilities of delay and average delays, respectively, at each stage. The distribution of total delays on delayed calls can be obtained, if necessary, by convoluting the separate distributions (Section 7.3). It is usually more convenient in practice, however, to specify a maximum delay at each stage, and the probability of exceeding that delay; the sum of these probabilities gives a measure of the total grade of service. At normal grades of service, in fact, the chance of a call suffering long delays at more than one stage is usually very small. While the assumption of independence between stages is strictly valid for negative exponential holding-time distribution only, a similar method may be applied generally, as a practical approximation.

In a telephone type of network, on the other hand, a call waiting for a stage-2 outlet continues to hold the preceding links in the connection, thus increasing the

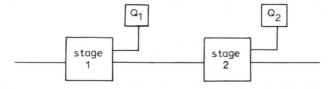

Fig. 47 Queues in series

traffic. This waiting traffic has a different distribution from normal traffic. Provided the waiting traffic is relatively small, however, as is usually the case with an acceptable grade of service, it can be allowed for approximately by increasing the traffic flow in erlangs on the links between stages 1 and 2, by the average number of waiting calls in Q_2 and any other queues succeeding stage 1, which can be calculated in the ordinary way (Section 7.2). This extra traffic should be allowed for in calculating the delays suffered by calls in Q_1, or the losses if stage 1 is a loss system.

7.10   Waiting systems with limited availability

We have seen that the equations of statistical equilibrium for a lost-call system with grading can be formulated in terms of the passage probability $u(x)$, i.e. the probability that a call is successful, given that the grading already contains x other calls (Section 4.7). In the case of a waiting-call system two functions are required, i.e. $u(x, s)$ = probability that a call is successful, given that the grading already contains x calls in progress and s calls waiting, $v(x, s)$ = probability that, after a call termination, the trunk is seized by a waiting call, given that there were x calls in progress and s waiting before the termination.

The general equations of statistical equilibrium are as follows: (Poisson input an d negative exponential holding times are assumed.)

$$[A + x]\, p(x, s) \quad = A\,[1 - u(x, s - 1)]\, p(x, s - 1) + Au(x - 1, s)p(x - 1, s)$$
$$+ (x + 1)[1 - v(x + 1, s)]\, p(x + 1, s)$$
$$+ xv(x, s + 1)p(x, s + 1)$$

$(0 \leqslant x \leqslant N, s \geqslant 0, N$ = total number of trunks in grading).

Thierer[101] gives a solution based on the following simplifying assumptions:

(*a*) That the grading is an Erlang ideal grading (Section 4.4)

(*b*) That the transition probabilities from state $(x - 1, s)$ to state (x, s) and from $(x, s - 1)$ to (x, s) are the same as those from (x, s) to $(x - 1, s)$ and from (x, s) to $(x, s - 1)$, respectively (Section 3.3).

By assumption (*a*), the k trunks available to a call, where k is the availability of the grading, form a random selection of k trunks out of N. The probability that they are all busy, given that the total number of busy trunks is x, is

$$1 - u(x, s) \;=\; \frac{\dbinom{x}{k}}{\dbinom{N}{k}}$$

which is independent of s, so that $u(x, s)$ may be written $u(x)$

Clearly
$$u(x) = 1 \quad \text{if} \quad 0 \leqslant x < k$$

By assumption (*b*)

$$x\{1 - v(x, s)\}p(x, s) = Au(x - 1, s)p(x - 1, s)$$

$$v(x, s)xp(x, s) = A\{1 - u(x, s - 1)\}p(x, s - 1)$$

Adding these equations we have

$$xp(x, s) = Au(x - 1)p(x - 1, s) + A\{1 - u(x)\}p(x, s - 1)$$

From this recurrence relationship, all *p*'s can be determined in terms of $p(0)$, which is itself determined by the condition that the sum of all *p*-terms is unity. This yields the following formulas:

The probability of a call being delayed is

$$p(>0) = \sum_{x=k}^{N} \left(\frac{\dfrac{A^x \prod\limits_{i=0}^{x-1} u(i)}{\prod\limits_{i=1}^{x} \{1 - A + Au(i)\}}}{1 + \sum\limits_{x=1}^{N} \left[\dfrac{A^x \prod\limits_{i=0}^{x-1} u(i)}{\prod\limits_{i=1}^{x} \{1 - A + Au(i)\}} \right]} \right)$$

The delay averaged over all calls is

$$W = \frac{\sum\limits_{x=k}^{N} p(x) \sum\limits_{i=k}^{N} \dfrac{A\{1 - u(i)\}}{1 - A + Au(i)}}{\sum\limits_{x=k}^{N} p(x)\{1 - u(x)\}}$$

Kuhn[40] has shown that these formulas can be applied to practical gradings, using an equivalent availability k' in place of k. For O'Dell gradings

$$k' = k - \frac{N^2 - k^2}{N^2} \left\{ \frac{3k}{4} \left(\frac{A}{N}\right)^{3/2} + \frac{ak}{10} \left(\frac{A}{N}\right)^2 \right\} \frac{1}{M-1}$$

where $M = gk/N$

g = number of grading groups

a = 0 for first-in-first-out service

a = 1 if different queues are served at random, the calls within any one queue being served in the order of arrival.

For gradings with skipping

$$k' = k - \frac{N^2 - k^2}{N^2} \left\{ \frac{kA}{5N} + \frac{k-3}{2} \left(\frac{A}{N}\right)^{k/4} + \frac{ak}{5} \left(\frac{A}{N}\right)^4 \right\} \frac{1}{M-1}$$

Kuhn has also determined the second moment of the delay distribution, and has shown that this distribution can be approximated by a gamma function.

7.11 Waiting system with conditional selection

If access to a route with delay facilities is obtained via a link system, the probability of congestion can be calculated by Jacobaeus's method in simple cases. Thus, in the scheme of Fig. 33, the distribution of busy links in the C column is calculated by Erlang's waiting-call formula, which, by a similar argument to that of Section 6.1, leads to the following expression for the probability of delay.

$$E = E_{2,\,m}(C)\left\{\frac{m - C}{mE_{1,\,m}(C/a)} + \frac{C}{m}\right\}$$

The effective availability method (Section 6.3) can also be used. The inlet blocking is determined as for a full-availability delay system. The average availability is calculated in the same way as for a loss system, and the route blocking can then be calculated as explained in Section 7.10.[41]

The delay averaged over all calls is the sum of the average delays for the inlets and outlets. The former is calculated as for a full-availability group, and the latter as for a grading with the same effective availability. The delay averaged over delayed calls is obtained by dividing the delay averaged over all calls by the total probability of blocking, which is

$$B_I = B_I + (1 - B_I)B_R$$

where B_I, B_R denote the inlet and route blocking, respectively.

7.12 Delays in stored-program-control systems

Stored-program-control (s.p.c.) systems, in which the switching of calls is controlled by a computer-type processor, give rise to delay problems of a special type.[123] In the first place, the operation of the processor is cyclic, and it may only be able to accept calls for service at a fixed point in the cycle . Secondly, processing of a call may not be a continuous operation. If processing is not completed within one cycle, the call may be returned to the queue, processing being continued the next time round. Two queue disciplines which are used in some s.p.c. systems are known as round robin (r.r.) and feedback (f.b.). In the former, there is only one queue, and an uncompleted call is returned to the end of it. In the latter, there may be any number of queues of different priority corresponding to different operations; e.g. scanning subscribers' lines for new call arrivals, route selection, operation of cross-points, testing for faults etc. When a call has received a 'quantum' of service, it is interrupted if there is a nonempty queue of higher priority.[102] Both these disciplines, particularly feedback tend to give better than average service to calls requiring shorter than average service time.[157]

Because of the high speed of the processor, the delays are usually too short to be noticed by subscribers, even when the processor occupancy approaches 100%, unless delay in the execution of a program results in call failure. Thus, the primary object of delay analysis is to determine the *maximum* traffic capacity of the processor under normal and fault conditions. A very slight reduction of traffic capacity below the maximum level is usually sufficient to satisfy any constraints necessary to ensure an acceptable grade of service to subscribers. The order of priority of different queues in the f.b. system can have an important influence on traffic capacity. To take an extreme example, if new calls were always served before uncompleted existing calls, the system could easily become blocked by uncompleted calls. The problem of optimising queue priorities is analogous to that of job scheduling in production control.[117]

Chapter 8

Nonblocking networks

Nonblocking networks are generally uneconomical with present technology but are of interest as a standard of reference with which practical networks can be compared. We will confine our attention to switching networks in which sources and sinks are identical, and any source/sink can be connected to any other; this is obviously the case when all sources are subscribers' lines. In practice, this is not strictly true, since a sink may be, for example, a speaking clock; moreover, the immediate sources and sinks of an exchange may include unidirectional junctions.

In the first instance we will consider the type of network in which each source/sink is connected to the network at two points; an inlet at which the calls it originates enter the network, and an outlet at which the calls it receives leave the network. The simplest nonblocking network is a square switch with N inlets, N outlets and N^2 crosspoints (Fig. 1a). It is not possible to improve on this with a 2-stage matrix. For suppose the first stage (A) comprises k switches with n inlets each, and the second stage (B) comprises m switches with p outlets each, so that $nk = mp = N$. Suppose $n \leqslant p$; it must be possible for all n inlets of any switch in the A stage to be connected simultaneously to any n outlets of any switch in the B stage. Hence, each A switch must have n links to each B switch. Each A switch, therefore, has n^2m crosspoints, and each B switch has knp crosspoints. The total number of crosspoints is

$$kn^2m + mknp = Nnm + N^2 > N^2$$

The same applies if $n > p$.

8.1 3-stage networks

Clos has shown that with three stages it *is* possible to design a nonblocking network with fewer than N^2 crosspoints.[30] A Clos-type network is shown in Fig. 48. The first stage (A) comprises k switches with n inlets each, serving the calling lines, and

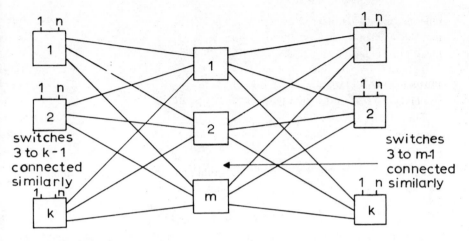

Fig. 48 Clos nonblocking 3-stage network

the last stage (C) comprises k switches with n outlets each, serving the called lines, so that $kn = N$, the total number of sources/sinks. The intermediate stage (B) comprises m switches with k inlets and outlets, so that each B switch has one link to each A and C switch.

Clearly m must not be less than n; otherwise blocking would occur when any A or C switch carried m calls, and one of the remaining n-m subscribers connected to that switch attempted a call. This, however, is insufficient to ensure the absence of blocking. Suppose a free inlet of switch A_1 requires connection to a free outlet of switch C_1. It is possible that the remaining $n - 1$ inlets of A_1 are already connected to outlets of switches other than C_1, and the remaining $n - 1$ outlets of C_1 are connected to inlets of switches other than A_1, so that $2n - 2$ B switches are unavailable to the new call. Hence, the minimum number of B switches to ensure absence of blocking is $m = 2n - 1$.

The minimum number of crosspoints is therefore

$$X_3(N, n) = 2kn(2n - 1) + mk^2$$

$$= (2n - 1)(2N + k^2)$$

$$= (2n - 1)N(2 + N/n^2)$$

For a given value of N, the number of crosspoints depends on the number of inlets per A switch n. The minimum value, not necessarily integral, is given by

$$\frac{\partial X_3(N, n)}{\partial n} = 0$$

thus

$$2n^3 - Nn + N = 0$$

thus

$$N = \frac{2n^3}{n-1} = 2n^2$$

approximately for large N.

Taking $n = (N/2)^{1/2}$, the total number of crosspoints is

$$X(N, (N/2)^{1/2}) = 4(2)^{1/2}N^{3/2} - 4N$$

$$= 2(2N)^{3/2} \text{ approximately for large } N.$$

It can be seen that the 3-stage network generally requires fewer crosspoints than the 1-stage network by taking the nonoptimal value $n = N^{1/2}$. Then

$$X(N, N^{1/2}) = 3N(2n - 1) = 6N^{3/2} - 3N$$

If there are values of N for which the single-stage network has the same number of crosspoints as, or fewer than the 3-stage network, these values must satisfy the inequality

$$N^2 \leqslant 6N^{3/2} - 3N$$

i.e. $N + 3 \leqslant 6N^{1/2}$

i.e. $N^2 - 30N + 9 \leqslant 0$

This holds only if $N < 30$, so the 3-stage network certainly requires fewer crosspoints for higher values of N at least, which includes the range of greatest practical interest.

Table 10 shows the approximate number of crosspoints for different values of N, with $n = (N/2)^{1/2}$ and $n = N^{1/2}$

Table 10

N	$X(N, (N/2)^{1/2}$	$X(N, N^{1/2})$
100	5327	5700
1000	174 884	186 738
10 000	5 616 800	9 970 000

8.2 General Clos-networks

A 5-stage nonblocking network can be obtained by replacing each B switch in Fig. 48 by a complete 3-stage nonblocking network. Let the five stages be designated A, B, C, D and E in consecutive order (Fig. 49), and let the A stage comprise k switches with n inlets and $2n - 1$ outlets as before. There are $2n - 1$ B-C-D networks, each with k inlets, one from each A switch. Let each of these 3-stage

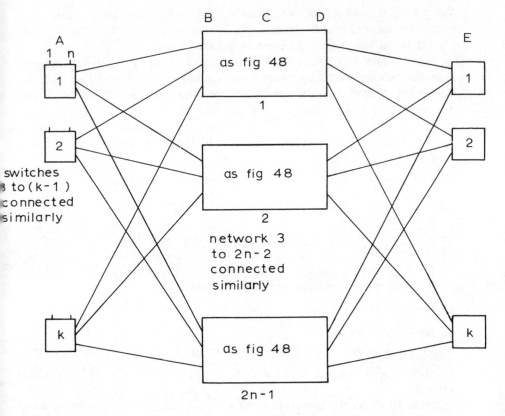

Fig. 49 Clos nonblocking 5-stage network

networks comprise k_1 B-switches with n_1 inlets and $2n_1 - 1$ outlets, so that $k_1 n_1 = k$. The D and E switches are symmetrical with the B and A switches, respectively, and each C-switch has k_1 inlets and outlets. Each B-C-D network has

$$(2n_1 - 1)k \left(2 + \frac{k}{n_1^2}\right) \text{ crosspoints,}$$

so the total number of crosspoints is

$$X_5(N, n, n_1) = 2kn(2n - 1) + (2n - 1)(2n_1 - 1)k \left(2 + \frac{k}{n_1^2}\right)$$

$$= (2n - 1)\left\{2N + (2n_1 - 1)\frac{N}{n}\left(2 + \frac{N}{nn_1^2}\right)\right\}$$

For a given value of N, the switch sizes required to minimise the number of crosspoints are obtained by differentiating $X_5(N, n, n_1)$ with respect to n and n_1, equating the partial derivatives to zero, and solving for n and n_1.

In the same way, a 7-stage nonblocking network can be constructed by replacing each C switch in the 5-stage network by a 3-stage network. Crosspoint minimisation involves three variables, so the problem of optimisation increases in complexity with the number of stages. However, a rough idea of the value of N at which it is advantageous to increase the number of stages can be obtained by assuming a fixed relation between N and the number of inlets per switch in the first stage n. A reasonable relationship is[17]

$$n = N^{2/(s+1)}$$

where s is the total number of stages.

A crosspoint comparison between this rule and the optimal rule for a 3-stage network has already been tabulated. In accordance with the general principle for construction of Clos networks, stages 2 to $s-1$ comprise $2n-1$ self-contained nonblocking networks of $s-2$ stages, each having $k = N/n$ inlets. The number of inlets per switch in the second stage is, by the same rule,

$$n_1 = k^{2/(s-2+1)}$$

and so on.

This leads to the following general formula for the number of crosspoints in an s-stage nonblocking network, in the form suggested by S.O. Rice and J. Riordan.[30]

$$X_s = \frac{n^2(2n-1)}{n-1}(5n-3)(2n-1)^{t-1} - 2n^t$$

with

$$t = \frac{s-1}{2}$$

$$n = N^{2/(s+1)}$$

As N increases, it is found that a 5-stage network becomes advantageous at about $N = 200$, and a 7-stage one at about $N = 5000$.

8.3 Triangular networks

Fig. 50 illustrates a type of nonblocking network suitable for a system in which each source/sink is connected to the network at one point only, which serves as both an inlet and an outlet. The crosspoints in stage B are only used to connect inlets in different A switches. It is assumed that two inlets of the same A switch can be connected via a pair of crosspoints in the A switch, thus busying a link to the B-stage but not affecting the crosspoints at that stage. For the network to be

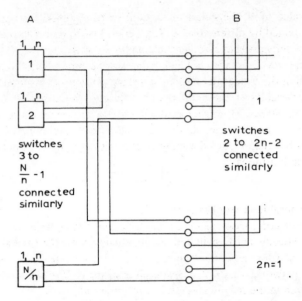

Fig. 50 Triangular nonblocking network

nonblocking, there must be at least $(2n - 1)$ B-switches as before. Each A switch has $n(2n - 1)$ crosspoints. Each B switch has N/n inlets, and a crosspoint corresponding to each possible pair of connected inlets, i.e.

$$\left(\frac{1}{2}\right)\left(\frac{N}{n}\right)\left(\frac{N}{n} - 1\right)$$

in all. The total number of crosspoints in the network is

$$X = \left(\frac{N}{n}\right)n(2n - 1) + \frac{(2n - 1)}{2}\left(\frac{N}{n}\right)\left(\frac{N}{n} - 1\right)$$

$$= (2n - 1)\left(N + \frac{N^2}{2n^2} - \frac{N}{2n}\right)$$

Minimisation of X with respect to n is carried out as before.

8.4 Other types of nonblocking network

Fig. 51A shows a rearrangeable 3-stage nonblocking network of the Clos type. Since it is always possible to rearrange existing calls so that free A-B and B-C links are connected to the same B switch, no allowance for internal blocking is

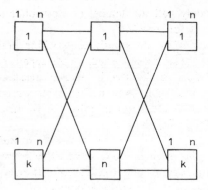

Fig. 51A Rearrangeable nonblocking network

necessary, so the A and C switches can be square. The total number of crosspoints is

$$X = 2kn^2 + nk^2 = 2(N/n)n^2 + n(N/n)^2 = 2Nn + N^2/n$$

The minimum value of X is given exactly by $n = (N/2)^{1/2}$, the same as the approximate value in the equivalent nonrearrangeable network. This corresponds to

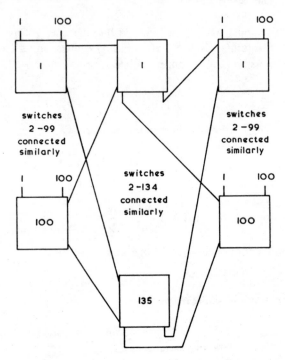

Fig. 51B Nearly nonblocking network

$X = (2N)^{3/2}$ which is only half the number of crosspoints required in the 3-stage nonrearrangeable network for large N.

All networks discussed so far are 'strictly nonblocking' in the sense that blocking cannot occur however the calls are routed. Other types are 'nonblocking in the wide sense', i.e. links must be allocated by certain rules to ensure the absence of blocking.[17] These may need less crosspoints than the equivalent strictly nonblocking networks. The design of networks of this type is difficult, and the principle has not found much practical application.

A more practical method of reducing crosspoint quantities is the 'quasi-' or 'nearly' nonblocking network, which produces very low internal blocking even when all inlets and outlets are fully loaded. This situation is highly unlikely to occur, so that the blocking with normal or even exceptionally high traffic is negligible.[31,32] An example is shown in Fig. 51B. Because the number of internal links in this 3-stage system considerably exceeds the number of inlets, the blocking probability is below 0·005 when every A-inlet and C-outlet carries 1 E. The total number of crosspoints is 4 050 000; a strictly nonblocking network with the same number of stages, inlets and outlets requires a minimum of about 5 600 000 crosspoints.[17]

The general traffic theory of rearrangeable networks has been investigated,[16] but the subject is of limited interest as, with present technology, rearrangeability is incompatible with an acceptable quality of transmission.

Chapter 9

Limitations of classical traffic and congestion models

As we have seen, Erlang and other pioneers of traffic theory made a number of simplified assumptions, which are still widely used as the basis of traffic engineering. In this Section we will consider the effect of these simplifications, and some more realistic theories which have been proposed.

9.1 Repeated attempts

Erlang's theory of loss systems assumes that calls which encounter congestion are lost, whereas, in practice, most unsuccessful callers make one or more further attempts to establish connection. Let us assume, in the first instance, that the intervals between successive attempts are very long, and that an unsuccessful call has a fixed probability of being repeated, however many previous attempts have been made. This assumption implies that the repeat attempt can be regarded as a new call, since the congestion state which produces it is too remote to affect the state of the system at the time of repetition.

Let p denote the probability of congestion, ϕ the probability that an unsuccessful call is repeated, and ψ the average number of attempts per call. The probability that a call is unsuccessful and is repeated at the first attempt is $p\phi$, and the probability that it is either successful at the first attempt or immediately abandoned is $1 - p\phi$. The probability that it makes just x attempts, in other words, that it is unsuccessful and repeated $x - 1$ times and is either successful or abandoned at the xth attempt, is $(p\phi)^{x-1}(1 - p\phi)$. Hence

$$\psi = \sum_{x=1}^{\infty} x(p\phi)^{x-1}(1 - p\phi) = \frac{1}{1 - p\phi}$$

Since assumptions (*a*) and (*b*) of Erlang's lost-call formula still apply (Section 3.1), the effect of repeat attempts is simply to increase the average arrival rate, and therefore the offered traffic, by a factor of ψ; the probability of congestion in a full-availability group of N trunks with A erlang offered (first attempt traffic) is therefore

$$p = \frac{\dfrac{(A\psi)^N}{N!}}{\displaystyle\sum_{r=0}^{\infty} \frac{(A\psi)^r}{r!}} = E_{1,N}(A\psi)$$

The general case where the repetition time is not necessarily long has been dealt with by Cohen[18] and a number of later authors who have made various assumptions regarding subscriber's behaviour. The following discussion is based on the work of Bretschneider.[156] It is assumed that, if a first attempt finds all trunks busy, it is repeated with probability ϕ_1 and lost with probability $1 - \phi_1$. For all subsequent attempts, the probabilities are ϕ_2 and $1 - \phi_2$. Repetition times have negative exponential distribution with mean values $1/s$. For computational purposes, it is necessary to assume that only a limited number of calls m can be waiting repetition simultaneously. Otherwise, the assumptions of Erlang's lost-call formula apply.

Let $p(x,y)$ = probability of x out of N trunks being busy with y calls waiting repetition and let (x,y) denote the state of question. Tables 11 and 12 show

(*i*)   the events which cause the system to leave the state (x,y) and their probabilities of occurence during a short time dt;

(*ii*)   the probabilities of all states which can be transformed to state (x,y) by a single event, and the probabilities of the events in question.

Note that a new call produces no change of state when $x = N$, if either the call is abandoned or the queue of calls waiting repetition is full ($y = m$); while a repetition produces no change if $x = N$ and the call is not abandoned but rejoins the queue. Equations of state are set up by equating the total probability of entering and leaving each state (x,y). These equations, together with the condition.

$$\sum_{x=0}^{N} \sum_{y=0}^{m} p(x,y) = 1$$

are sufficient to determine all the $p(x,y)$ terms.

The probability of all trunks being busy is $\sum_{y=0}^{m} p(N,y)$; this is the time congestion. It is also the call congestion in respect of new calls since their arrival is independent of the state of the system with Poisson input, but not in respect of repeat calls, except in the limiting case of very long repetition times. The traffic carried is

$$\sum_{x=1}^{N} \sum_{y=0}^{m} xp(x,y),$$

so the probability of loss is

$$1 - \frac{\sum_{x=1}^{N} \sum_{y=0}^{m} xp(x,y)}{A}$$

Table 11   Events causing system to leave state (x, y)

x	y	Events	Probability during dt	New state
$x = 0$	$y = 0$	new call	$Ap(0, 0)dt$	$(1, 0)$
$0 < x < N$	$y = 0$	new call	$Ap(x, 0)dt$	$(x + 1, 0)$
		termination	$xp(x, 0)dt$	$(x - 1, 0)$
$x = N$	$y = 0$	new call (not abandoned)	$A\phi_1 p(N, 0)dt$	$(N, 1)$
		termination	$Np(N, 0)dt$	$(N - 1, 0)$
$x = 0$	$0 < y \leqslant m$	new call	$Ap(0, y)dt$	$(1, y)$
		repetition	$ysp(0, y)dt$	$(1, y - 1)$
$0 < x < N$	$0 < y \leqslant m$	new call	$Ap(x, y)dt$	$(x + 1, y)$
		repetition	$ysp(x, y)dt$	$(x + 1, y - 1)$
		termination	$xp(x, y)dt$	$(x - 1, y)$
$x = N$	$0 < y < m$	new call (not abandoned	$A\phi_1 p(N, y)dt$	$(N, y + 1)$
		repetition (abandoned)	$ys(1 - \phi_2) p(N, y)dt$	$(N, y - 1)$
		termination	$Np(N, y)dt$	$(N - 1, y)$
$x = N$	$y = m$	repetition (abandoned)	$ms(1 - \phi_2) p(N, m)dt$	$(N, m - 1)$
		termination	$Np(N, m)dt$	$(N - 1, m)$

An alternative expression for the probability of loss can be obtained as follows: The probability that interval dt contains a new call arrival is Adt, while the probability that it contains a new or repeated call which is unsuccessful and abandoned (or finds the queue full) is

$$\left\{ A(1 - \phi_1) \sum_{y=0}^{m} p(N,y) + A\phi_1 p(N,m) + \sum_{y=1}^{m} ys(1 - \phi_2) \ p(N, y) \right\} dt$$

The ratio of the latter expression to Adt is the ratio of losses to new calls over a long period, in other words, the probability of loss.

Pure loss and delay systems can both be regarded as limiting cases of a system with repeated attempts. The former corresponds to zero probability of repetition

Table 12   Events causing system to enter state (x, y)

x	y	Previous state	Event	Probability during dt
$x = 0$	$0 \leqslant y \leqslant m$	$(1, y)$	termination	$1p(1, y)dt$
$0 < x < N$	$0 \leqslant y < m$	$(x - 1, y)$	new call	$Ap(x - 1, y)dt$
		$(x - 1, y + 1)$	repetition	$(y + 1)sp(x - 1,$ $y + 1)dt$
		$(x + 1, y)$	termination	$(x + 1)$ $p(x + 1, y)dt$
	$y = m$	$(x - 1, m)$	new call	$Ap(x - 1, m)dt$
		$(x + 1, m)$	termination	$(x + 1)$ $p(x + 1, m)dt$
$x = N$	$y = 0$	$(N - 1, 0)$	new call	$Ap(N - 1, 0)dt$
		$(N - 1, 1)$	repetition	$1sp(N - 1, 1)dt$
	$0 \leqslant y < m$	$(N - 1, y)$	new call	$Ap(N - 1, y)dt$
		$(N - 1, y + 1)$	repetition	$(y + 1)$ $sp(N - 1, y + 1)dt$
		$(N, y - 1)$	new call (not abandoned)	$A\phi_1 p(N, y - 1)dt$
		$(N, y + 1)$	Repetition (abandoned)	$(y + 1)s(1 - \phi_2)$ $p(N, y + 1)dt$
	$y = m$	$(N - 1, m)$	new call	$Ap(N - 1, m)dt$
		$(N, m - 1)$	new call (not abandoned)	$A\phi_1 p(N, m - 1)dt$

$(\phi_1 = \phi_2 = 0)$, and the latter to certainty of repetition with infinitesimal intervals between attempts and no limit to the number of calls waiting repetition at one time $(\phi_1 = \phi_2 = 1, s = \infty, m = \infty)$. In these cases the probability of congestion becomes $E_{1,N}(A)$ and $E_{2,N}(A)$, respectively, while the probability of loss becomes $E_{1,N}(A)$ and zero. If $\phi_1 = \phi_2 = 1$ but s is finite and m infinite, the probability of loss is zero; as with the pure delay system, equilibrium is not attained if $A \geqslant N$.

If the repetition probability is less than 1, it is clear that a very short average interval between attempts would not reduce the losses very much as compared with a pure loss system, since most repeat attempts would be made before the original state of congestion had cleared. However, if the average interval is of the same order as, or larger than, the average duration of a congestion state ($1/N$ for a full-availability group), as is usually the case in practice, its magnitude has relatively little effect on loss probability. The most important parameters, therefore, are the repetition probabilities ϕ_1 and ϕ_2. Typical values quoted by Bretschneider are

$$\phi_1 = 0.6, \qquad \phi_2 = 0.75$$

He gives loss and congestion probabilities for these data in a 10-trunk group with 4 E and 8 E offered. The loss probability is reduced by about 50% as compared with

Erlang's lost-call formula, while the congestion probability (time congestion) is increased by less than 10%. As this example illustrates, the effect of repeat attempts is not sufficiently serious to invalidate the normal basis of equipment provision for lost-call systems. The probability of blocking at the first attempt, which is generally adopted as a measure of grade of service, is equal to the time congestion.

Repeated calls are generated not only by congestion but also by called lines being engaged. Since the busy states of different lines, however, are relatively independent they are not closely correlated with system congestion, except in special circumstances, such as when a high proportion of traffic on a trunk group is directed to a large and overloaded p.b.x. group. Hence, the theoretical importance of this type of repetition traffic is relatively small.

In fact, the practical importance of repeat attempts of all kinds lies not so much in their effect on the distribution law of call arrivals in a congested group as in the extra load they impose on other parts of the network, particularly on common controls and under fault conditions (Chapters 11, 12).

9.2 Waiting-call defections

The commonly used waiting-call formulas are based on the assumption that a caller is prepared to wait for service indefinitely, whereas, in practice, long delays cause a certain proportion of calls to be abandoned. The effect is usually negligible with normal traffic, since, given adequate provision of equipment, the chance of a call being delayed to the limit of a subscriber's patience is very small. Call defection may, however, be significant under heavy overload conditions, due either to abnormal demand or faults.

With call defection, the grade of service experienced by a particular subscriber depends both on his own reaction to delay and on that of other subscribers. The longer subscribers as a whole are prepared to wait, the fewer calls will be abandoned, and, therefore, the more traffic will be carried. As a result, however, time congestion is significantly increased, so that a relatively impatient subscriber is more likely to abandon a call than he would be if most others were equally impatient, since the route would then be less congested.

We will assume that both the holding time and the time for which subscribers are prepared to wait have negative exponential holding-time distribution, and that queue discipline is first-in-first-out. Since calls which cannot be served are eventually abandoned, a state of equilibrium is attained, even if the traffic offered, in erlangs, exceeds the number of trunks.

Let A be the traffic offered to a full-availability group of N trunks, and let $1/L$ be the average time which a subscriber is prepared to wait for service, expressed in terms of the average holding time. If there are j waiting calls, the probability of one being abandoned during a short time dt is $jL dt$, while the probability of an existing call terminating is $N dt$, as usual. Denoting the probability of there being x calls in the system by $p(x)$, we obtain the following equations of equilibrium (assuming Poisson input)

$$Ap(x) = (x + 1) p(x + 1) \qquad (0 \leqslant x < N)$$

$$Ap(x) = \{ N + (x + 1 - N)L \} p(x + 1) \qquad \geqslant N$$

The solution is

$$p(x) = p(0) \frac{A^x}{x!} \qquad (0 \leqslant x < N)$$

$$p(x) = p(0) \frac{A^x}{N!} \{(N + L)(N + 2L) \dots (N + (x - N))L\}^{-1} \qquad (x > N)$$

$$p(0) = \left\{ \sum_{r=0}^{N-1} \frac{A^r}{r!} + \frac{A^N}{N!} K \right\}^{-1}$$

where

$$K = \frac{(N/L)e^{A/L}\gamma(N/L, A/L)}{(A/L)^{N/L}}$$

$$\gamma(N/L, A/L) = \int_0^{A/L} x^{(N/L)-1} e^{-x} dx$$

(the incomplete gamma function)

The following important results can be proved:

the probability that all N channels are busy

$$E_D = \frac{KE_{1,N}(A)}{1 + (K - 1)E_{1,N}(A)}$$

the probability that a call does not wait for connection

$$P = \frac{N}{A} \left(\frac{1}{K} + \frac{A}{N} - 1 \right) E_D$$

the waiting time averaged over all calls, including abandoned calls

$$\bar{w} = A \frac{N}{L} \left(\frac{1}{K} + \frac{A}{N} - 1 \right) E_D$$

the probability of a delay exceeding t, assuming the caller is prepared to wait at least t

$$p(>t) = \frac{E_D \gamma(N/L, Ae^{-t/L})}{\gamma(N/L, A/L)}$$

The values $L = \infty$, and $L = 0$, of course, correspond to a loss system and a delay systems without defection; it can be shown that the foregoing expressions take the appropriate limiting values in these cases.

These formulas were first derived by Palm.[28] They are applicable when the system treats part of the traffic on a loss basis and part on a delay basis. An example of this situation is a bothway concentrator connecting a group of subscribers to the exchange. Outward calls wait for dialling tone but inward calls receive busy tone in the event of congestion. The only modification in the formulas is the substitution of a for A, except in the term $E_{1,N}(A)$, which is left unaltered (a is the traffic subject to delay).

Observation of dialling delays by Clos and Wilkinson[104] were in reasonably good agreement with Palm's theory. The average time for which callers were prepared to wait varied from 24 s to 74 s, depending on the class of line. Wikell's observations in Sweden gave an average waiting time of 100–110 s; the proportion of abandoned calls was somewhat less that that predicted by the formula.[106, 105]

Call defections may have a significant effect on the loading of manual operators, enquiry office staff etc.[105] As an example, consider a group of 10 operators, the traffic offered being 8 E. If the average holding time (i.e. operating time in this context) is 30 s, and the average for which subscribers are prepared to wait is 60 s, the probability of delay is 0·32, as compared with 0·41 in the absence of call defection. The corresponding figures for the delay averaged over all calls, including abandoned calls, are 2·3 s and 6·1 s. The proportion of abandoned calls is 3·9%, which reduces the average operators' occupied time from 80% to 76·9%.

If subscribers are prepared to wait, on the average, for one holding time, so that $L = 1$, Palm's formula for the probability of all trunks being busy reduces to Molina's lost-calls-held formula (Section 3.7). This may seem paradoxical, since Molina assumes that the caller waits for a fixed time, and then, if successful, holds the trunk for the unelapsed duration only. Molina's formula is valid, however, for any distribution of waiting time. Moreover, in the case of negative exponential holding-time distribution, the unelapsed duration of held calls has the same distribution as the total duration of all calls, so the result is to be expected.

Call defections can be allowed for in a similar manner in the case of a limited number of calling sources. The following formula[29] gives the probability of blocking (call congestion) for M sources, each of which, when free, originates traffic at a rate a. In the limit, as $L \to \infty$, it is equivalent to Engset's lost-call formula (Section 3.5).

$$B = \frac{\displaystyle\sum_{x=N}^{M-1} \{(M-1)!a^x(N/L)!\}/\{(M-1-x)!N!L^{x-N}(N/L+x-N)!\}}{\displaystyle\sum_{x=0}^{N-1} \{(M-1)!a^x\}/\{(M-1-x)!x!\} + \sum_{x=N}^{M-1} \{(M-1)!a^x(N/L)!\}/ \{(M-1-x)!N!L^{x-n}(N/L+x-N)!\}}$$

9.3 Day-to-day traffic variations

Owing to variations in the average traffic during different busy hours, a succession of busy hours is not equivalent to a continuous equilibrium process, and the average blocking is usually greater than equilibrium theory predicts. There are several methods of allowing for this. A common method is to stipulate the permissible blocking for specified percentage increases in traffic flow above the normal value (Section 2.8). This does not, however, allow for the fact that some routes may have more variable traffic than others, and may suffer relatively worse average grade of service, since the overload limits are likely to be reached with greater than average frequency. Uniformity of service on different routes is particularly important for international traffic. The International Consultative Committee on Telecommunications (CCITT) has studied traffic variations on international routes, and has made the following recommendation as a result.[107] The loss probability for the mean busy hour traffic of the 30 highest days should not exceed 1%, while that for the mean busy-hour traffic of the five highest days should not exceed 7%. This rule provides better protection against excessive degradation of service on particular routes than a simple percentage-overload criterion.

These methods generally employ traffic tables based on the classical equilibrium theory of Erlang. In other words, equilibrium is assumed for the purpose of calculation, but departures from it are allowed for by basing provision of trunks on a higher-than-average busy-hour traffic. An alternative approach is to aim at a specified average grade of service for the busy *season*, the grade being fixed at such a level that deviations during particular busy hours can be tolerated. It has been found empirically that Molina's lost-calls-held formula (Section 3.7) gives a better estimate of congestion than Erlang's lost-calls-cleared formula when the traffic is averaged over a large number of busy hours.[58] Further improvement can be obtained by using one of the call-defection formulas described in Section 9.2 to calculate the probability of blocking; in this application, the parameter L is not necessarily regarded as a measure of waiting time, but simply as an empirical constant, the value of which depends on the properties of the traffic on the route in question, whether the system operates on a loss or delay basis.

A more rigorous approach is to calculate the average congestion during the busy season as the weighted mean of the busy-hour congestion values, taking account of traffic variations. If A denotes the traffic offered to a route, and $f(A)dA$ is the probability that, during a particular busy hour, its value lies between A and $A + dA$, the average probability of blocking during the season is

$$\int_{A=0}^{\infty} f(A)B(A)dA$$

where $B(A) =$ probability of blocking for A erlang in a state of equilibrium, calculated from Erlang's, or another appropriate formula.

Traffic tables have been calculated on this basis, assuming A to have a normal or gamma distribution; the latter avoids the possible anomaly of negative loads.[108-110] The combined effect of day-to-day and within-busy-hour (random) variations is conveniently expressed in terms of the ratio of the variance of the number of simulraneous calls to its mean. For very stable traffic the effect of day-to-day variation may be negligible, so that the distribution is practically Poissonian, the variance/ mean ratio being unity. Observations in the USA indicate that the ratio varies from 1 to about 1·9, the highest figure being applicable to final routes in alternative routing networks (Section 11.3); this allows for the 'peakiness' of the traffic offered as well as the for day-to-day variations in average load.[57]

The effect of variations in traffic intensity *within* the busy hour, as distinct from random fluctuations around a constant mean value, was discussed by Palm,[138] but has not found much application in traffic engineering. Modern measurement techniques, however, make it possible to determine such variations more accurately than was the case in the past,[139] and it may well be desirable to take account of them in the design of s.p.c. systems, where short departures from equilibrium may have a considerable effect on the call-handling capacity of a high-occupancy processor.

9.4 Unbalanced traffic in small groups of subscribers

Engset's formula for congestion with a limited number of traffic sources assumes that all sources originate the same average load. The effect of uneven source loading for a given total traffic is to reduce the probability of blocking as is easily seen in an extreme case. If a small proportion of sources carry a very high proportion of the traffic, the effective number of sources is obviously less than the actual number, which reduces the probability of high traffic peaks. On the other hand, a low traffic source suffers a relatively poor grade of service, since the competing traffic from other sources is greater than in the case of a high traffic source. The theory of congestion with unevenly loaded sources has been investigated by Cohen[111] and Dartois.[112]

A more important problem from the practical point of view is the difficulty of distributing the sources over different switching groups so that all groups carry the same average traffic. While gross unbalances can usually be avoided by spreading known heavily used lines evenly, exact balance is impracticable. The traffic in each group can, however, be predicted within specified confidence limits if the frequency distribution of subscribers' demand is known. Let a and s denote the mean and standard deviation of the traffic on a subscriber's line; these may refer to originated, terminated, or total traffic, depending on the function of the trunks under consideration. Then the traffic on a random selection of N lines has mean value Na and a variance of Ns^2. Plausible distribution functions for the traffic per line, allowing for the impossibility of negative values, are negative exponential, beta and gamma.[113,114] In the first case mentioned, the standard deviation is equal to the mean. Note that an extreme upper bound for the variance of the traffic on N lines is $Na(1-a)$. This corresponds to a loading of 1 E per line in a proportion a of the lines, and 0 E in the remainder.

A reasonable margin of safety might be achieved by provisioning for an estimated traffic load of 1·65 standard deviations above the mean, irrespective of the distribution law. This limit would be exceeded with probability 0·05 if the sample mean were normally distributed, which may be the case approximately even when the distribution is nonnormal. As an example, suppose the mean traffic per line, averaged over the whole exchange area, is 0·06 E, and its standard deviation is 0·08. The traffic on a random sample of 100 lines has mean 6 E and standard deviation of 0·8. A reasonably safe maximum traffic estimate is therefore

$$6 + 1·65 \times 0·8 = 7·3 \, E$$

In practice, given reasonable traffic balancing with regard to known high-demand lines, grouping is not completely random, which provides a further margin of safety.

Chapter 10

Problems involving connection time

It is usually assumed in traffic theory that seizure of any traffic-carrying device is instantaneous, since the connection time is normally too small to make any significant difference to traffic calculations. There are certain problems, however, where it has to be taken into account. For example, there may be a risk of double connection if two calls attempt to seize the same device within a short period, and the system must be designed to reduce this risk to acceptable levels. Again, the average number of outlets tested by a switch before finding a free one may affect the wear and, therefore, the life of the equipment. Excessive searching time may entail loss of fast-dialled calls. The discussion that follows is based mainly on the work of Fry.[52]

10.1 Hunting by finders

If a finder switch (Section 1.8) serves n lines, and all are equally loaded, the average number of outlets traversed, including the outlet which requires connection, is the mean of the maximum and minimum, i.e. $(1 + n)/2$. The probability that the required outlet is not among the first k traversed, or, in other words, the probability of having to traverse more than a specified number of outlets k is

$$1 - (k/n)$$

These formulas hold whether or not the finder returns to a home position when released, and whether the multiple is straight or slipped.

If nonhoming switches are employed, the average connection time can be reduced by permitting a number m of finders to search for the same source simultaneously. It may be assumed that the finders are positioned independently, so the probability that all of them test more than k outlets is

$$p(>k) = \left(1 - \frac{k}{n}\right)^m$$

The probability that a successful switch tests exactly k outlets (i.e. more than $k-1$ but not more than k) is

$$p(k) = p(>k-1) - p(>k)$$

The average number of outlets tested by the successful switch is

$$\sum_{k=0}^{n} kp(k) = 1\{p(>0) - p(>1)\} + 2p\{(>1) - p(>2)\}...$$

$$... + (n-1)\{p(>n-2) - p(>n-1)\} + np(>n-1)$$

$$= \sum_{k=0}^{n-1} p(>k) = n^{-m} \sum_{k=0}^{n-1} (n-k)^m$$

$$= n^{-m} \sum_{k=1}^{n} k^m$$

10.2 Hunting by selectors

Consider a full-availability group of N trunks with A erlang offered, lost-calls cleared and sequential hunting. If all trunks are busy, it is assumed that the switch returns to its home position and the caller receives busy tone. The probability of testing more than k trunks is $E_{1,k}(A)$ (Section 3.4). The probability of testing exactly k trunks is therefore

$$E_{1,k-1}(A) - E_{1,k}(A)$$

The average number of trunks tested is

$$\sum_{k=0}^{N-1} E_{1,k}(A)$$

With random hunting or slipped multiple, the probability that the first k outlets to be tested are all busy is given by the Palm–Jacobaeus formula (for proof see Section 4.6), i.e.

$$p(>k) = \frac{E_{1,N}(A)}{E_{1,N-k}(A)}$$

The average number of trunks tested is therefore

$$\sum_{k=0}^{N-1} p(>k) = E_{1,N}(A) \sum_{k=0}^{N-1} \{E_{1,N-k}(A)^{-1}\}$$

10.3 Double connection

Suppose there is a danger period d while a call is being connected to a trunk, during which period the same trunk may be picked up by a second call. Assuming sequential hunting, the conditions for double connection are as follows:

(*a*) another call is offered within a time *d* after the arrival of the first call;

(*b*) none of the earlier choices are released in time to intercept the second call.

If the switch hunts at the rate of *s* outlets per unit time, the maximum number of trunks tested by the second call in time *d* is *sd*. If *d* is very short, as it must be for tolerable service, the probability that any of these is released in this time is approximately *sd*/*h*, where *h* is the average holding time. Taking account of the fact that even release during *d* may not be in time to intercept the call, the probability of condition (*b*) occuring is at least $(1 - sd/h)$.

The average number of calls arriving in time *d* is *Ad*/*h*. Hence, for the Poisson input, the probability of condition (*a*) occuring is

$$1 - e^{-A/dh} \doteqdot Ad/h$$

From the point of view of a caller who is being connected to a trunk, the probability that his call will suffer interference from a later call is approximately

$$Ad/h\{1 - sd/h\} \doteqdot Ad/h$$

Clearly, any call is equally likely to interfere with a later call as to be interfered with by an earlier call. The total probability that a call is involved in a double connection is therefore

$$2Ad/h - \phi \doteqdot 2Ad/h$$

where ϕ is the comparatively negligible probability that the same call both interferes with an earlier call and suffers interference from a later call.

In the case of random hunting, the second call has probability $1/(N - x)$ of picking up the same trunk as the first call, where *x* is the number of calls in progress, excluding these two; the probability that other calls may also be in the danger period is neglected. Assuming lost calls are cleared, *x* has an Erlang distribution. Strictly speaking, allowance should be made for the fact that at least one trunk is known to have been free when the first call arrived, i.e. the one it has just seized. The probability of exactly *x* calls being in the system at that moment, i.e. the conditional probability of *x* calls, knowing that the system contains less than *N* calls, is given by

$$p(x \mid < N) = p(x)/p(<N)$$

where $p(x)$ and $p(<N)$ denote the unconditional probabilities of exactly *x* and less than *N* calls respectively.

Hence

$$p(x \mid < N) = \frac{A^x/x!}{\sum_{r=0}^{N-1} A^r/r!}$$

The probability that a call suffers interference from a later call is therefore

$$\sum_{x=0}^{N-1} \frac{Adh}{N-x} p(x|{<}N)$$

The total probability that a call suffers double connection is approximately twice this value, which reduces to

$$\frac{2d\left(\displaystyle\sum_{x=0}^{N-1} \frac{A^{x+1}}{(N-x)x!}\right)}{h \displaystyle\sum_{r=0}^{N-1} A^{r}/r!} \equiv 2f(A)d/h$$

Fry ignores the difference between $p(x)$ and $p(x|{<}N)$, but the error is negligible at normal grades of service. He has evaluated $f(A)$ for a range of values of N and A/N (Reference 52).

Example

Find the probability of double connection, with sequential and random hunting, with $N = 10$, $A = 4{\cdot}0E$, $d = 100\,\text{ms}$ and $h = 100\,\text{s}$.

The required probability is

$$2 \times 4 \times 0.001 = 0.008$$

$$2 \times 0{\cdot}75 \times 0{\cdot}001 = 0{\cdot}0016 \quad \text{(random, using Fry's curves)}$$

If there is a danger of a new call picking up the same trunk as a previous call while the latter is being cleared down, and thereby preventing the release of the previous connection, similar formulas apply, with a slight difference in the conditional probability distribution. With sequential hunting, a necessary condition for double connection is that all the earlier choices should be busy when the caller begins to clear. The probability that the clearing call is on the xth choice is the ratio of the traffic carried by that choice to the total traffic carried, i.e.

$$\frac{A\{E_{1,x-1}(A) - E_{1,x}(A)\}}{A\{1 - E_{1,N}(A)\}}$$

The conditional probability that the first $x - 1$ choices are busy, knowing that the xth choice is busy, is given by

$$\psi(1, x - 1|x) = \psi(1, x)/\psi(x) \quad (1 < x \leqslant N)$$

where

$\psi(1, x) = $ unconditional probability of choices 1 to x inclusive being busy

$\psi(x) = $ unconditional probability of choice x being busy, irrespective of the state of any other choices.

Hence

$$\psi(x) = \text{traffic carried by choice number } x.$$

Hence

$$\psi(1, x - 1|x) = \frac{E_{1,x}(A)}{\{E_{1,x-1}(A) - E_{1,x}(A)\}A}$$

The probability that the clearing call is interfered with by a later call is therefore

$$Ad/h \sum_{x=1}^{N} \left(\frac{A\{E_{1,x-1}(A) - E_{1,x}(A)\}}{A\{1 - E_{1,N}(A)\}} \right) \left(\frac{E_{1,x}(A)}{\{E_{1,x-1}(A) - E_{1,x}(A)\}A} \right)$$

$$= \frac{Ad}{h\{1 - E_{1,N}(A)\}} \sum_{x=1}^{N} E_{1,x}(A)$$

(It is permissible to extend the range of x to include $x = 1$, since the expression for $\psi(1, x - 1|x)$ then becomes 1; this may be interpreted as meaning that, if the clearing call is on the first choice, a call arriving within the danger period will cause a double connection with probability 1).

The above formula also gives the probability that a call will make double connection with an earlier call which is just clearing down.

In the case of random hunting, the second call has probability $1/(N-x+1)$ of picking up the same trunk as the clearing call, where x is the number of calls in progress, *including* the clearing call. As before, the probability that any other call is also clearing down at the same time is neglected. The conditional probability of there being x calls, given that there is at least one call, is

$$p(x|\geqslant 1) = \frac{A^x/x!}{\sum_{r=1}^{N} A^r/r!}$$

The probability that a clearing call suffers interference from a later call is therefore

$$\frac{Ad}{h} \left(\frac{\sum_{x=1}^{N} A^x/\{(N-x+1)x!\}}{\sum_{r=1}^{N} A^r/r!} \right)$$

Chapter 11

Traffic aspects of network planning

In planning a telecommunication network, the objective is to determine the numbers, sizes, locations and boundaries of exchanges, and the arrangements and quantities of junctions between them, in such a way as to minimise total cast, subject to meeting the stipulated standards of performance (including grade of service, reliability and quality of transmission). The mathematical theory of networks has developed rapidly in recent years, and many algorithms are known for connecting a given number of points with minimum cost, finding the shortest route between two points etc. These algorithms can sometimes be of value in the planning of telecommunication networks.[115,116] Generally speaking, however, the number of feasible alternative networks is too large for complete analysis. The usual practical procedure is to select a reasonably efficient network as a starting point and attempt to improve it by successive modifications. There are generally quite a number of acceptable alternatives with little difference in cost. Thus, by repeating the calculation with a number of different initial networks, one can be reasonably confident of finding a solution which is near the true optimum. It may be helpful to plot the cost of successive trials, or whatever other objective function is being minimised, as shown in Fig. 52. A smooth curve drawn through the lowest points found up to, and including, each trial tends to level off as the optimum is approached. This method, which is sometimes known as the Las Vegas technique,[117] is applicable to purely random trials as well to systematic optimum-seeking procedures, but the latter, of course, are likely to require fewer trials.

11.1 Distribution of exchanges

Each exchange, however small, entails a substantial fixed cost, so, the fewer exchanges that are provided, the lower the total cost of the buildings and equipment. On the other hand, to reduce cabling costs, the average distance between the subscriber's instrument and the exchange should be as small as possible,which demands

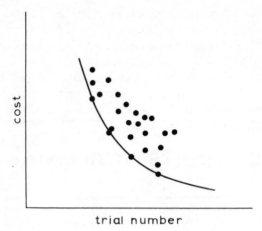

Fig. 52 Optimisation assessment

a large number of exchanges. The problem can be formulated mathematically as follows: For simplicity, the transmission network between exchanges is omitted; it must of course be included in a complete formulation. For fuller details see Anderberg *et al.*[118] Let the area be divided into rectangles of a convenient size by means of a grid of I rows and J columns.

Let

$s(i, j)$ = number of subscribers in rectangle (i, j)

$d(i, j, r)$ = mean distance of a subscriber in rectangle (i, j) from exchange r

E = total number of exchanges

$h(i, j, r)$ = 1 or 0 according to whether subscribers in rectangle (i, j) do or do not belong to exchange r

k = fixed cost per exchange

$c\{d(i j r)\}$ = cost of a subscriber's line of length $d(i, j, r)$

c may be a step function, because transmission specifications may require cable with lower attenuation and/or resistance over a certain distance.

The total cost, so far as this depends on the distribution of exchanges, is

$$C = Ek + \sum_{r=1}^{E} \sum_{i=1}^{I} \sum_{j=1}^{J} s(i, j) h(i, j, r) c\{d(i j r)\}$$

The cost of exchange equipment which is in proportion to the number of lines or the traffic is omitted, as this is assumed to be fixed in total, irrespective of the number of exchanges and their distribution.

The problem is to choose values of E, exchange locations (which determine the $d(i, j, r)$ terms, and exchange boundaries (which determine the $h(i, j, r)$ terms so as to minimise C. A feasible procedure is as follows:

(*a*) Start with a trial number of exchanges, and locate them by judgement.

(*b*) Optimise the exchange boundaries.

(*c*) Optimise the exchange locations (in terms of grid references).

(*d*) Repeat (*b*) and (*c*) until stability is reached.

(*e*) Search for locations at the intersections of exchange boundaries where the installation of another exchange would reduce total cost.

(*f*) Place exchanges in these locations and reoptimise all boundaries and locations.

(*g*) Repeat, deleting exchanges if necessary until no further cost reduction is obtained.

11.2 Structure of transmission network

The next step is to determine the layout of transmission routes between exchanges. In the case of a very small network a 'mesh' pattern may be used, every exchange being connected to every other (Fig. 53*a*). The number of direct unidrectional routes with N exchanges is $N(N-1)$. Thus, as N increases, the number of routes in a mesh increases very rapidly, and the average traffic per route becomes very small, which is inefficient (Chapter 3). At the opposite extreme, the minimum number of routes connecting N exchanges is $(N-1)$. There are a large number of possible minimum-route networks ('trees' in graph theoretical terminology). It is often convenient to arrange the exchanges in a hierarchy, each one being connected to a number of others in the next lower rank (Fig. 53*c*); a 2-rank hierarchy is

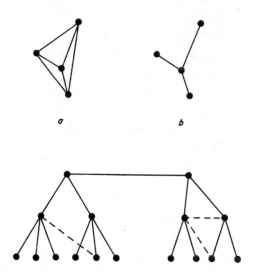

a   *b*

c

Fig. 53 Network patterns

called a 'star network' (Fig. 53b). This pattern facilitates standardisation, since the various control and switching functions can be associated with specific levels in the hierarchy. The second and higher rank exchanges may be purely transit centres, switching calls between other exchanges, or they may also serve local subscribers, possibly be means of a separate switching unit in the same building.

In addition to the basic hierarchical network, auxiliary direct routes may also be provided between certain pairs of exchanges, in the same or different ranks, as shown by dotted lines in Fig. 53c. This is justified when the traffic on the direct route is large enough to ensure that the cost of providing the direct route is outweighed by the saving in circuits in the indirect route. A lower grade of service may be used in direct routes because calls using them generally require a lower total number of links than those in the basic network. The following cost factors should be taken into account in deciding whether or not to provide a direct route,[119] for the purpose of calculation all items are expressed in terms of annual charges.

(a) Line costs, including cable, repeater equipment and terminal equipment associated with the junctions.

(b) Switching costs at the terminal and intermediate switching points.

(c) Rearrangement costs arising from the provision of a new direct route.

As an example, suppose traffic between two exchanges X and Z is at present routed via an intermediate exchange Y, but the provision of a direct XZ links is contemplated.

Let

N = number of trunks in the direct route

n_1, n_2 = number of trunks saved in links XY and YZ, respectively, by the introduction of direct route XZ; these quantities may be unequal because of differences in availability etc. of the two links.

K = total annual charges associated with the direct route, including items (a) (b) and (c); K may be of the form $G + HN$, where G and H are constants depending on the length and type of route.

k_1, k_2 = annual charges per trunk saved, in links XY and YZ, respectively, as a result of the provision of a direct XZ link, including items (a) and (b).

Provision of the direct route is justified if

$$K \leqslant n_1 k_1 + n_2 k_2$$

Trunks between exchanges may be either one-way or bothway. The latter require extra equipment, but this may be justified if there is insufficient traffic in one direction to load a separate trunk group efficiently, or if the directional distribution of traffic fluctuates to such an extent that separate groups would be uneconomical.

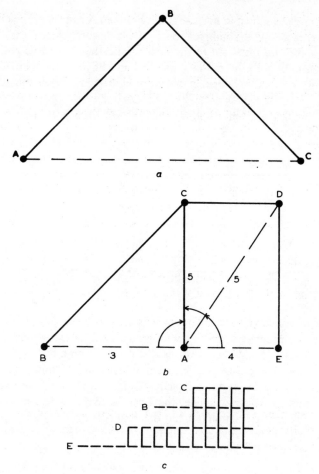

Fig. 54 Alternative routing

Sometimes a mixture of early choice unidirectional and late choice bothway trunks is provided. This arrangement is equivalent to a 2-group progressive grading, the commons representing bothway trunks. The traffic capacity can be calculated by the equivalent random method (Section 4.5).

11.3 Alternative routing

In some systems, a call is only allowed to take one route, which may be in either the basic or the auxiliary network, or partly in both. In that case, all routes must have sufficient trunks to give an adequate grade of service. Other systems employ alternative routing (the American usage 'alternate routing' is now widely adopted).

A simple example is shown in Fig. 54*a*. Calls between A and C have the direct route AC as first choice and the indirect route ABC as second choice; calls between A and B or B and C are restricted to the direct routes between those points. So long as sufficient trunks are provided in routes AB and BC to carry the overflow from AC, a high probability of blocking can be tolerated in AC, so this route can be provided on a high-usage basis.

A more complex example is shown in Fig. 54*b*. The possible routes for calls originating at A are

Table 13

To	1st choice	2nd choice	3rd choice
B	AB	ACB	
C	AC		
D	AD	ACD	
E	AE	ADE	ACDE

The number of trunks in each link from A is shown in Fig. 54b; the arrangement is equivalent to a progressive grading with groups of different availability as shown in Fig. 54*c*. Routes AB, AD and AE may be provided on a high-usage basis, with a high proportion of overflow traffic, but the final route AC must contain a sufficient number of trunks to ensure an adequate grade of service to all classes of call using it. In this particular example, the final route also serves as a first choice route to calls for one destination (C), but this is not always the case.

The mean *a* and the variance of the total traffic offered to the route AC, and the overflow *f* from AC, can be calculated by the equivalent random method. The overall probability of loss on this route is f/a. The parcels of traffic offered to AC from different early choice routes may not all have exactly the same probability of loss on AC. The differences, however, are to some extent self-compensating. Traffic destined for E, after overflowing its 1st- and 2nd-choice routes, will have a relatively high degree of 'peakiness' and will therefore suffer a relatively high proportional loss. However, it can afford to do so, since this loss applies, not to the original traffic but only to the part not carried by earlier choice routes. On the other hand, traffic destined for C, which has AC as its only route, is Poissonian, and suffers a relatively low loss, as it should do for an acceptable grade of service.

A route which serves as the final route for any class of traffic must be provided with sufficient trunks to ensure an acceptable grade of service to that traffic, taking account of any previous choices available to it. It is usually sufficiently accurate for this purpose to assume that all parcels of traffic to the route suffer the same probability of loss. The error is usually on the safe side, in the sense that the critical loss, for the purpose of dimensioning the route, is that suffered by traffic for which it is an early choice; the method tends to over-estimate this loss, leading, if anything to slight over-provision. The greatest error occurs when, as in this example, some traffic has access to one route only, which it shares with over-flow traffic from other routes. In this case, a better estimate of loss can be obtained, if necessary, by the following formula, due to Brockmeyer.[120]

Consider a full-availability group of $S + C$ trunks with sequential hunting. The probability that the last C trunks are all busy, irrespective of the stage of the first S trunks, is

$$R_1 = E_{1,S+c}(A) \frac{\sum\limits_{m=0}^{S} \binom{C+m}{m} \dfrac{A^{S-m}}{(S-m)!}}{\sum\limits_{m=0}^{S} \binom{C-1+m}{m} \dfrac{A^{S-m}}{(S-m)!}}$$

where A = traffic offered in erlangs.

Hence, if C is the number of trunks in AC, and S is the number of trunks in a full-availability group which, when offered a suitable amount of Poisson traffic, gives an overflow with the same mean and variance as the total traffic offered to route AC, then R_1 is the probability of loss for the first choice traffic offered to AC.

If the traffic destined for C is not too small, it may be economical to reserve a few trunks in route AC, either early or late choices, exclusively for this traffic. Alternatively, this could be given priority when the number of free trunks falls to a certain value. This may avoid over-dimensioning the route for the benefit of a small proportion of traffic.[121]

In the equivalent random method, the number of trunks in the equivalent group and the hypothetical traffic offered are interdependent. To facilitate computation, however, Rapp has devised the following approximate formula which gives the traffic independently of the number of trunks:

$$Y = V + (3V/M)\{(V/M) - 1\}$$

where M, V are the mean and variance of the total traffic offered to a final or inter-mediate route, and Y is the hypothetical traffic which, if offered to a full-avail-ability group, would produce overflow traffic with parameters M, V.[151] The num-ber of trunks in this full-availability group is given by

$$N = [Y/\{1 - 1/(M + V/M)\}] - M - 1$$

As an example, suppose that in Fig. 54a, 10 E originates at A and terminates at B, while 20 E originates at A and terminates at C; that the number of trunks from A to C is 10; and that the stipulated probability of congestion for route AB is 0·02. All trunks are assumed to be unidirectional. The traffic offered to route AB com-prises Poisson traffic with mean and variance 10, and overflow traffic with mean and variance given by

$$20 E_{1,10}(20) = 20 \times 0·538 = 10·76 E$$

and

$$V = 10 \cdot 76 \left(1 - 10 \cdot 76 + \frac{20}{11 - 20 + 10 \cdot 76} \right) = 17 \cdot 2$$

The mean and variance of the total traffic offered to route AB are

$$M = 10 + 10 \cdot 76 = 20 \cdot 76$$

and

$$V = 10 + 17 \cdot 2 = 27 \cdot 2$$

thus

$$Y = 27 \cdot 2 + \frac{3 \times 27 \cdot 2}{20 \cdot 76} \left(\frac{27 \cdot 2}{20 \cdot 76} - 1 \right) = 28 \cdot 4 \, \text{E}$$

$$N = \frac{28 \cdot 4}{1 - 1/(20 \cdot 76 + 1 \cdot 31)} - 20 \cdot 76 - 1 = 8 \cdot 0$$

In other words, $28 \cdot 4$ E Poisson traffic offered to a full-availability group of $8 \cdot 0$ trunks would produce overflow traffic with the same parameters as that offered to route AB.

The permissible overflow from route AB is

$$20 \cdot 76 \times 0 \cdot 02 = 0 \cdot 415 \, \text{E}.$$

If the required number of trunks is N_{AB}, the number in the equivalent full-availability group is

$$8 \cdot 0 + N_{AB}$$

Hence $28 \cdot 4$ E, offered to a full-availability group of $8 \cdot 0 + N_{AB}$, must produce an overflow not exceeding $0 \cdot 415$ E, so that

$$E_{1, 8 \cdot 0 + N_{AB}}(28 \cdot 4) \leqslant 0 \cdot 415 / 28 \cdot 4 = 0 \cdot 015$$

whence $8 \cdot 0 + N_{AB} \geqslant 39, N_{AB} \geqslant 31$ $E_1,$

11.4 Optimisation of alternative routing networks

The criteria for a decision, whether or not to provide an alternative direct route between two points, are essentially the same as for the provision of a direct route without overflow facilities (Section 11.2). First of all, however, it is necessary to determine the optimum number of trunks in order to cost the route. The following analysis is due to Pratt.[122]

To begin with, consider the simple triangular network of Fig. 54*a*. Calls from A destined for C use the high-usage route AC as first choice, and the final route ABC as second choice, Calls from A destined for B, and calls from B destined for C have no alternative route. All trunks are assumed to be unidirectional, and those in the other direction are not shown; generally, the two networks would be similar.

Let

N_{AC} = number of trunks from A to C

C_{AC} = cost of one trunk from A to C

A_{AC} = traffic offered to route AC

a_{AC} = traffic overflowing from route AC to route ABC

$E_{AC} = a_{AC}/A_{AC}$ = call congestion on route AC.

It is assumed that E_{AB} and E_{BC} are fixed, independently of any changes in N_{AC}, by the grades of service required by traffic for which these are first choice routes or by other requirements. Given A_{AC} and N_{AC} the value of a_{AC} and its variance can be calculated (Section 4.5). Moreover

$$A_{AB} = a_{AC} + \text{traffic from A destined for B.}$$

The second term on the right-hand side is known from the traffic data, and is assumed to be Poissonian. Hence, the mean and variance of A_{AB} can be calculated. The same applies to A_{BC}. Knowing the characteristics of the traffic offered to each link in the final route, the number of trunks therein, N_{AB} or N_{BC}, depends only on the required grade of service. The cost of the network for any trial value of N_{AC} can therefore be determined; the problem is to choose N_{AC} in such a way as to minimise the total cost

$$C = C_{AC}N_{AC} + C_{AB}N_{AB} + C_{BC}N_{BC}$$

As an approximation, the number of trunks may be treated as if it were a continuous quantity, in which case the necessary condition for a minimum value of C is

$$\left(\frac{\partial C}{\partial N_{AC}}\right) = 0$$

with A_{AC}, E_{AB} and E_{BC} being held constant,

i.e.

$$0 = C_{AC} + C_{AB}\left(\frac{\partial N_{AB}}{\partial N_{AC}}\right) + C_{BC}\left(\frac{\partial N_{BC}}{\partial N_{AC}}\right)$$

$$= C_{AC} + C_{AB}\left(\frac{\partial N_{AB}}{\partial a_{AC}}\right)\left(\frac{\partial a_{AC}}{\partial N_{AC}}\right) + C_{BC}\left(\frac{\partial N_{BC}}{\partial a_{AC}}\right)\left(\frac{\partial a_{AC}}{\partial N_{AC}}\right)$$

Since the overflow from AC is the only variable part of the traffic offered to routes AB and BC, $(\partial N_{AB}/\partial a_{AC})^{-1}$ and $(\partial N_{AC}/\partial a_{BC})^{-1}$ are the marginal traffic capacities of those routes; i.e. the increments in offered traffic per unit increment in trunks, required to keep the grades of service at the fixed values E_{AB} and E_{BC}. $-(\partial a_{AC}/\partial N_{AC})$ is the marginal occupancy of route AC; i.e. the increment in traffic carried per unit increment in trunks, the offered traffic being kept constant. Putting

$$\beta_{AB} = \left(\frac{\partial N_{AB}}{\partial a_{AC}}\right)^{-1} , \beta_{BC} = \left(\frac{\partial N_{BC}}{\partial a_{AC}}\right)^{-1} , H_{AC} = -\left(\frac{\partial a_{AC}}{\partial N_{AC}}\right)$$

the minimum-cost equation can be written

$$C_{AC}/H_{AC} = C_{AB}/\beta_{AB} + C_{BC}/\beta_{BC}$$

H_{AC} is the slope of the curve obtained by plotting the traffic carried $(A_{AC} - a_{AC})$ against N_{AC}, with A_{AC} held constant; this curve can be derived from the general form of Erlang's lost-call formula, which allows for nonintegral values of N (Appendix 1.8). β_{AB} is the slope of the curve obtained by plotting A_{AB} against N_{AB}. A_{AB} is changed only by changing the component a_{AC}, any other parcels of traffic offered to route AB being held constant; and a_{AC} is changed by changing N_{AC} while A_{AC} is held constant.

The value of N_{AB} for each value of A_{AB} at the specified grade of service E_{AB} can be determined by the equivalent random method, to allow for the nonPoisson component of A_{AB}.

The optimum number of trunks in each route can now be determined.

(i) Guess the values of β_{AB} and β_{BC}.

(ii) Calculate H_{AC} from the minimum-cost equation (see above).

(iii) Read N_{AC} from graphs of N against A with H as a parameter.

(iv) Calculate a_{AC} by Erlang's formula, then N_{AB} and N_{BC} by the equivalent random method.

(v) Calculate β_{AB} and β_{BC} as explained above. If these differ from the assumed values by more than a specified amount, repeat steps (i)–(v) until stability is obtained.

Similar procedures can be applied to more complex networks. If the network contains k high-usage or intermediate groups, there are k optimising equations of the form

$$\left(\frac{\partial C}{\partial N_r}\right) = 0$$

where C is the total cost of the network and N_r is the number of trunks in route r; these equations can be solved to obtain optimum values of N_r. As before, the numbers of trunks in final routes are dependent on the N_r terms, the traffic distribution and the specified grade of service.

11.5 Overall network control (network management)

Modern control techniques make it possible, although not necessarily economical in all circumstances, to extend the principle of conditional selection to a whole communication network, so that interexchange links and internal switching matrixes form a single-integrated system with end-to-end path selection. This could be effected either by centrally located control equipment or by interexchange signalling.

While this technique would increase the normal traffic capacity of a network at the specified grade of service, it would thereby render it more sensitive to overloading. This is true even for a single group of trunks (Chapter 3); in a network, the effect is aggravated by the fact that, as congestion approaches, calls have to find increasingly devious routes to get through, thereby occupying more trunks. The solution to this dilemma may lie in dynamic or adaptive control, i.e. changing the routing rules according to the state of the system.[43] This philosophy recognises that the value of sophisticated traffic techniques lies in ensuring that purely local peaks do not cause excessive blocking while other possible routes are idle. When general overloading arises, these techniques can do little to help and may even make matters worse. There is therefore a case for a systematic reduction of alternative routing facilities when congestion reaches a dangerous level. Moreover, there is unlikely to be any significant advantage in developing more elaborate procedures than the type of alternative-routing schemes which have already been described. Given sufficient information about the state of the network, and complete flexibility in the choice of route, it would no doubt be possible to ensure that some calls were completed which would otherwise be lost. The facility would, however, have to be curtailed or withdrawn just when it was most in demand.

This does not apply so much to message-switching systems of the store-and-forward type, in which a message occupies only one link at a time, the preceding link being released when the call advances to the next stage. In these circumstances, complex routing strategies may be advantageous even with heavy traffic, and this possibility may influence the optimum network structure.

The simplest situation to control arises when the traffic distribution alters in a regular pattern. For example, if a network comprises three exchanges X, Y and Z, and X serves a business area while Y and Z serve neighbouring residential areas, then routes XY and XZ would probably have their busy hours in the morning or afternoon, and YZ in the evening. Thus, calls between Y and Z might be alternatively routed via X in the evening, while, during the daytime, Y and Z might each serve as an alternative routing point between X and the other. Rerouting could be carried out at a fixed time, and would not require any indication of traffic conditions.

When a purely local overload occurs due to abnormal traffic within a small part of the network, which may also form part of alternative routes between remote exchanges, it may be possible to reduce congestion by suspending these alternative routes, thus diverting traffic from the affected area. Thus, in Fig. 54*a*, if there is heavy overloading in exchange B, calls from A to C could be temporarily restricted to the direct route.

If the overloading is due to abnormally heavy traffic coming into the area from outside, this obviously cannot be diverted away, and traffic restriction may be desirable. The first step could be to remove alternative-routing facilities from calls destined for the affected area. This in itself would reduce the incoming load by causing blocking. If this is insufficient, some trunks in the direct route might be busied out, or bothway trunks made unidirectional.

Other steps which may help to reduce congestion during short overloads include giving priority to the completion of calls which have nearly reached their destination, and delaying the receipt of dialling tone for a minute or more, except for essential services. The latter is most effective in systems in which calling lines can be prevented from occupying exchange equipment; this is easily arranged in s.p.c. exchanges, by suspending the scanning of calling lines until the equipment is ready to deal with new calls.

To apply these techniques effectively, early detection of overloading is essential. Convenient indications of network congestion include the number of incoming calls awaiting service by registers, the proportion of unsuccessful attempts, the occupancy of final routes etc. Overload controls may be imposed either by judgment or automatically in the light of this information. The derivation of decision rules, however, is a complex problem, and at present there is a lack of useful algorithms applicable to a variety of networks.

Chapter 12

Traffic measurement

Regular traffic measurement is essential in order to check the adequacy of equipment provision and to obtain reliable data as a basis for planning the growth of the system. The cost of collecting and analysing traffic data in a large network is high, but is justified by the benefits of improved service and of plant savings which are made possible by accurate traffic estimates.

12.1 Measurement of traffic flow

The most important parameter to be measured is the busy-hour traffic flow in each route, which determines the provision of equipment. It would be possible to estimate the flow from the number of calls, which is easily measured by call-counting meters, and the average holding time. Direct determination of the latter, however, would require regular timing of large samples of calls, while the possibility of correlation between holding time and calling rate is another complication. In practice, the most convenient method of measuring traffic flow is to scan each traffic-carrying device at regular intervals and record whether it is free or busy. The number of 'busies' divided by the number of times the device was scanned gives an estimate of the proportion of time for which it was occupied, which is equal to the traffic carried in erlangs. The total traffic on a group of devices is obtained by dividing the total number of 'busies' by the number of complete scanning cycles. Scanning can be performed visually but automatic traffic recording equipment is generally employed to save labour.[34] Large exchanges incorporate permanent traffic recorders, while portable recorders are available for small exchanges.

The recorder is normally arranged so that the traffic on each full-availability group, grading etc. is recorded on a separate meter. At the same time, the connections between the traffic carrying devices and the recording meters are flexible, so that the traffic on partial groups or single devices can be recorded if necessary.

The scanning interval is usually in the range 30 s to 3 min for speech-carrying equipment and 6—18 s for short-holding-time equipment, such as common controls. In the case of very short holding-time equipment such as markers, however, scanning may not be the most suitable method of measurement; the holding time may be known accurately from the operating characteristics of the system, in which case, the traffic flow can be obtained from call-counting meters. The accuracy of the scanning method is dealt with later.

Ideally, it should be possible to measure the traffic on all trunk groups simultaneously. To economise in recording equipment, however, the system may be arranged so that the exchange is divided into sections (normally three), each of which is measured separately.

12.2 Busy-hour traffic

Since the busy hour may vary, it is necessary to record the traffic during a longer period than 60 min to ensure that the busy-hour traffic is measured. One method is to record the traffic during, say, four successive half-hour intervals, covering three overlapping hours. The highest hourly traffic flows on a number of representative days are averaged to obtain the busy-hour traffic. This is known as the post-selected busy-hour method.

An alternative is the time-consistent busy-hour method, in which the traffic during the same hour on different days is averaged, the fixed hour with the highest average traffic being adopted as the busy hour. This method is more convenient than the other, since, once the busy hour has been determined by preliminary measurements, the recorder can be automatically switched on during this hour for as many days as necessary, and the meters need only be read at the beginning and end of the record.

If the true traffic intensity is constant during the measurement period, so that variations are due to random causes, the time-consistent method gives an unbiased estimate of traffic flow, and is therefore theoretically sounder than the post-selected method. The former naturally tends to give lower traffic estimates than the latter, typically by 5% to 10%.[36] If the method of measurement is changed, this difference can be compensated for by adjusting the nominal grade of service used for provisioning.

Apart from routine measurements of the traffic on each trunk group, special measurements may be necessary to check that the traffic on certain groups is properly distributed. For example, unbalanced loading of grading groups may produce excessive congestion even though the total traffic is well within the theoretical capacity of the grading. This can be checked by comparing the traffic on the individual choices serving different groups.

12.3 Switch-count error

Traffic recording is essentially a sampling process, and is subject to two distinct types of error. First, the traffic flow is observed, not continuously, but only at fixed intervals. Secondly, the traffic during a particular busy hour is a sample from a population of busy hours, so that even if it were measured accurately, the resulting estimate of the true traffic intensity would be subject to error. These types will be designated 'switch-count error' and 'traffic-estimation error', respectively.

If the holding time is constant and equal to or greater than the scanning interval, the switch-count error is virtually zero. Every call is counted the same number of times, except for a vanishingly small proportion of calls, the end-points of which happen to coincide with scanning times; this can only happen if the holding time is an exact multiple of the scanning interval.

If the holding time is constant and less than the scanning interval no call is ever scanned more than once, so that the number of calls recorded during successive scans are independent, if calls occur at random. For Poisson input, neglecting the effect of congestion, the variance of the number of simultaneous calls is equal to the mean, i.e. the traffic flow in erlangs (a). The variance of the mean number of simultaneous calls derived from a sample of Z/i scans is the ratio of the population variance to the sample size, i.e. ai/Z, where a is the actual traffic carried (Z = total time of observation). This formula is applicable to other holding-time distributions, provided the scanning interval is long compared with the average holding time, or exceeds the maximum holding time when this has a definite upper limit, as might be the case with control equipment.

We have assumed, in effect, that all trunks are scanned simultaneously. In practice the time lag is too small to affect the error significantly, but it can be seen to have theoretical significance by postulating a very slow scanner. In that case, the condition of each trunk at the moment of scanning would be independent of the state of the previous trunk when it was scanned a long time before. Assuming all trunks are equally loaded, the number of calls recorded in one scan has binomial distribution with mean a and variance $a(1 - a/N)$, where N is the number of trunks. With simultaneous scanning, however, the variance of the number of calls recorded at a single scan is a as we have seen.

For the general theory of switch-error count, allowing for correlation between successive scans, see Hayward.[35] The variance of the estimated traffic flow is approximately

$$\sigma_1^2 = \left\{ r \left(\frac{1 + e^{-r}}{1 - e^{-r}} \right) - 2 \right\} ha/Z$$

where $r = i/h$
h = average holding time
a = traffic carried in erlangs
i = interval between successive scans
Z = total time of observation

This formula assumes Poisson input, negative exponential holding time distribution, and negligible congestion. The approximation is good provided

$$Z/h > 20$$

If r is large, $\sigma_1^2 \doteq rha/Z = ai/Z$ as before.

12.4 Traffic-estimation error and total error

Neglecting congestion, the average number of calls entering the system during the observation period Z is AZ/h where A is the traffic *offered*, and the average value of their total duration is AZ. If Z is long compared with h, this expression is the average value of the traffic volume U (Section 2.1) offered during Z, since the effect of calls in progress at the beginning and end of the period is negligible. The variance of U, allowing for variations in calling rate and holding time, is (Appendix 1.7)

$$AZ(\sigma_h^2 + h^2)/h$$

where $\sigma_h^2 =$ variance of holding time.
The traffic flow during Z is U/Z and its variance is

$$\sigma_2^2 = \frac{AZ}{hZ^2}(\sigma_h^2 + h^2) = \frac{A(\sigma_h^2 + h^2)}{hZ}$$

This becomes Ah/Z and $2Ah/Z$ in the case of constant and exponential holding time, respectively.

The total variance of the measured traffic, allowing for switch-count error and estimation error, is

$$\sigma^2 = \sigma_1^2 + \sigma_2^2$$

Neglecting congestion, and assuming negative exponential distribution, we have

$$A = a, \sigma_h^2 = h^2$$

$$\sigma^2 \doteq r\left(\frac{1 + e^{-r}}{1 - e^{-r}}\right)hA/Z = \left(\frac{1 + e^{-r}}{1 - e^{-r}}\right)iA/Z$$

12.5 Measurement of congestion

Measurement of carried traffic does not provide a very sensitive indication of congestion, since a large increase in offered traffic may produce very little increase in trunk occupancy if this is already high. Traffic recorders are therefore supplemented by meters which measure congestion more directly.

In the case of a full-availability group, the time congestion can be measured by means of a circuit which connects a meter to a time pulse whenever all trunks are busy simultaneously. If i is the interval between pulses, Z the observation period, and m the meter reading, the time congestion is $B = mi/Z$.

Since congestion is only recorded at fixed intervals, there is a sampling error which can be calculated in the same way as the switch-count error in traffic recording. The average duration of a congestion state, in a group of N trunks with A erlang offered,

is $\qquad T = h/N$ for a loss system

or $\qquad T = h/(N - A)$ for a waiting system.

where h is the average holding time.

The variance of the measured time congestion is

$$\sigma^2 = \left\{ i/T \left(\frac{1 + e^{-i/T}}{1 - e^{-i/T}} \right) - 2 \right\} TB/Z$$

It may be useful to measure the total number of congestion states, as well as the total time, to distinguish between frequent and isolated blocking, by means of a separate counting meter. If the average holding time is known, an estimate of time congestion can be obtained from this meter alone, independently of the time pulse. If m is the number of congestion states occuring during a time Z, the estimated time congestion is mT/Z.

The same method of measuring congestion is applicable to gradings, but separate meters are required for each grading group. In the case of progressive gradings, it is more convenient to measure the time during which the last choice is occupied, and/ or the number of times it is seized. The last-choice traffic corresponding to a specified grade of service can be calculated approximately by the following method:

Let N = total number of trunks in the grading

k = availability

A = traffic offered in erlangs

B = probability of congestion calculated from the appropriate formula (e.g. equivalent-random or m.P.J.)

The expected overflow traffic is AB erlang.

Let A_1 = traffic which, when offered to a full-availability group of k trunks, gives overflow AB.

A_k = traffic offered to the kth trunk in a sequentially hunted full-availability group when A_1 erlang is offered to the first trunk.

Thus $\qquad AB/A_1 = E_{1,k}(A_1)$

$$A_k = A_1 E_{1,k-1}(A)$$

A_1 and A_k can be determined with the help of curves showing the traffic offered to successive choices in a full-availability group, calculated from Erlang's formula. The traffic carried by the last choice in the grading is approximately $A_k - AB$. If the meter indicates a higher value, there is evidence of overloading which should be checked by a traffic record. The last choice load can be derived more accurately by the equivalent random method. This, however, is hardly necessary, since high accuracy is not required to indicate serious overloading, which is the primary purpose of congestion meters. For the same reason, it is unnecessary to treat a delayed-call grading differently from a lost-call grading in this respect.

Overflow meters, which record the number of calls lost in each trunk group owing to congestion, are applicable to any type of loss system. They are, however, less accurate than time-congestion meters, because they are subject to error arising from random variations in calling rate as well as in the duration of congestion.[152] Moreover, the number of overflows may be greatly influenced by repeated attempts by unsuccessful callers, which is difficult to allow for without special observation. The expected number of overflows during a period Z, ignoring repeated attempts, is AZB/h. If the meter reading is consistently higher than this, there is evidence of overloading. The effect of repeated attempts is considered later.

12.6 Average-holding-time measurement

The average holding time could, of course, be measured by timing a random sample of calls. This may require a rather large sample. Assuming negative exponential distribution, the standard deviation σ_h is equal to the mean h. The standard deviation of the mean of samples of N calls is

$$h/\sqrt{N}$$

Suppose we require to determine the mean within $\pm 10\%$ accuracy with 95% confidence. The distribution of the mean of a large sample is approximately normal. In a normal distribution, 95% of the population is contained within a range of ± 1.96 standard deviations from the mean. The minimum sample size is therefore given by

$$\frac{1.96h}{\sqrt{N}} = 0.1h$$

thus
$$N = 385$$

If call-counting meters are equiped, the average holding time can be calculated from the total number of calls (or, rather, attempted calls) C and the measured traffic flow A during a time Z. The average holding time per attempt is, of course, AZ/C.

12.7 Allowance for lost traffic

For most traffic calculations, a knowledge of the offered rather than the carried traffic is required; to derive this information from traffic-flow measurements, it is necessary to estimate the lost traffic. This can be done by using the appropriate congestion formula. Thus, in the case of α full-availability group of N trunks with Poisson input and lost calls cleared, we have

$$A = a/\{1 - E_N(A)\}$$

where a and A denote the carried and offered traffic, respectively. A is not an explicit function of a, but can either be computed iteratively or read from a graph. In fact, with normal loading, the difference may be small enough to ignore; or A can be replaced by a on the right-hand side of the above equation, leading to an explicit function of A which can be quickly evaluated with the aid of Erlang loss tables. Under overload conditions, however, any serious departure from the assumptions on which the formula is based will obviously reduce the accuracy of the loss estimate derived from it. In particular, the effect of repeated attempts may be significant; this can be allowed for as follows:

Let r = proportion of successful *attempts*;
$\quad \beta$ = average number of attempts per call ($\beta \geqslant 1$)

During one average holding time, the average number of first attempts is equal to the traffic offered A so the average total number of attempts is βA, of which $\beta A r$ are successful. Hence $\beta A r = a$, the traffic carried.

Thus A can be calculated from the measured value of a if r and β are known. r could be measured by equipping each trunk group with call-counting and overflow meters. β could be determined by special observation; it is generally a function of r. since, the higher the congestion, the more repeat attempts are necessary.

To avoid the complexity of a rigorous theoretical model for repeat attempts (Section 9.1), it may be sufficiently accurate to treat them as independent events; in which case, the system behaves like a lost-calls-cleared system with βA erlang offered. This equivalent offered traffic might be estimated from the carried traffic by using Erlang's formula, as already explained, and used for dimensioning the equipment, without actually knowing the value of β. If the existing equipment is overloaded, this would lead to overprovision of new equipment, since the repeat-attempt traffic would be reduced.

A further complication is that the traffic carried at one stage may be reduced as a result of congestion at earlier stages, which may prevent calls from reaching it, or by congestion at a later stage, which may lead to premature release. Thus, provision of extra trunks or common controls at a congested stage may lead to increased traffic flow at other stages, producing more congestion if this has not been allowed for. It is therefore necessary to consider the effect of congestion on the traffic circulation in the network as a whole, taking account of repeat attempts. The following simplified discussion of the subject is largely based on the work of Le Gall.[19]

Consider the traffic flowing into and out of any section of a network. This 'section' may be a single trunk group, a set of connected trunk groups forming a step-by-step or link system, or even the complete network. The inlets and outlets of the system may be calling and called subscribers' lines, and/or intermediate trunks. In general, we can distinguish between the original source traffic intended for the section under consideration, and the traffic offered by the section inlets, since there may be losses at earlier stages. This does not, of course, apply if the inlets are the original calling sources. In order that the concept of traffic offered at the section inlets shall be meaningful, it is assumed that the section and its inlets do not form part of a single conditional-selection system; such a system must be considered as a single 'section' for our purpose.

Let A_F = average number of first attempts per average *conversation*-holding time

B_1 = probability that an attempt is blocked before reaching the section inlets

B_2 = probability that an attempt, having seized a section inlet, is blocked from entering the section

B_3 = probability that an attempt which succeeds in entering the section is blocked within the section or at a later switching stage. If the section comprises a single stage, or a number of conditionally selected stages arranged so that a call cannot enter the section unless there is a through path, this term refers to later stages only

q = probability that the called subscriber is free and the attempt is answered

r_1, r_2, r_3 = proportions of attempts intended for, offered to, and entering the section, respectively, which are effective, i.e. connected to the required subscriber and answered

$\beta(r_1)$ = average number of originated attempts per call when the proportion of effective attempts is r_1

θ = uncharged holding time per attempt, averaged over all attempts which enter the section, whether effective or not (expressed as a proportion of the average conversation time).

Ineffective attempts include those encountering congestion or called-line engaged, or receiving no answer; those due to faults, however, will be neglected. We have

$$r_1 = (1 - B_1)r_2$$
$$r_2 = (1 - B_2)r_3$$
$$r_3 = (1 - B_3)q$$

The effective traffic carried by the section is, by definition, the average number of effective attempts per average conversation holding time, which is $A_F\beta(r_1)r_1$. The average number of all attempts, successful or not, which enter the section during the average conversation-holding time, is

$$A_F\beta(r_1)(1 - B_1)(1 - B_2)$$

The uncharged traffic carried by the section is therefore .

$$A_F\beta(r_1)(1 - B_1)(1 - B_2)\theta = A_F\beta(r_1)r_1\theta/r_3$$

The total traffic carried by the section is

$$a = A_F\beta(r_1)r_1(1 + \theta/r_3)$$

If all stages in the existing network are adequately provisioned, congestion at any stage may have negligible effect elsewhere, in which case it is safe to engineer network extensions on a stage-by-stage basis. The equivalent 'lost-calls-cleared' traffic offered to the stage under consideration in the existing network is

$$a/(1 - B_2) = ar_3/r_2$$

which can be calculated if either r_2 and r_3 or B_2 can be measured. Assuming that these remain unaltered after extension, the new equivalent offered traffic, which is used to dimension the stage under consideration in conjunction with the appropriate lost-calls-cleared traffic table, can be obtained simply by applying a growth factor, if necessary, to allow for increased demand.

If, however, there is serious overloading in the network, it may be necessary to allow for the effect on general traffic circulation when congested stages are alleviated. To begin with, let us consider the network as a whole, the inlets and outlets under consideration being calling and called lines. Then

$$B_1 = 0$$

$$r_1 = r_2 = (1 - B_2)r_3 = (1 - B_2)(1 - B_3)q$$

Whichever of the quantities r_1, r_3, B_2, B_3 and q are most conveniently measurable can be used to determine the others. If necessary, theoretical values of B_2 and B_3 may be used when available measurements are insufficient. If it is possible to distinguish charged and uncharged traffic in measuring traffic flow, θ can be obtained directly as the ratio of the latter to the former, If not, it can be estimated approximately, using average values of setting-up time, ringing time, time of listening to busy tone and conversation time.

Using the measured carried traffic a and the empirical function $\beta(r_1)$ we can estimate the original source traffic A_F for the existing network. A growth factor is applied to obtain the value of A_F for the new network. If the form of $\beta(r_1)$ is not known, it may be sufficiently accurate to calculate the product $A_F\beta(r_1)$ for the present network and apply the growth factor to that.

The equivalent offered traffic for provisioning each stage can be determined in the same way.

12.8 Computerised traffic measurement

Computerised systems of traffic measurement have recently been developed in which traffic data are recorded directly on magnetic tape for processing.[44] This reduces the labour involved in taking and analysing the records and increases their scope. For example, it is possible to record the traffic all day and analyse its variations so that shifts in busy hour can be detected. By transmitting the information over data links to a central computer continuous monitoring of traffic variations over a whole network becomes a practical possibility. This is ideally desirable for traffic management purposes, but was not feasible with older techniques.

12.9 Allocation of traffic to subscriber classes

For planning purposes it is frequently necessary to know the average originated or terminated traffic per line for different classes of subscribers, such as residential, business (single-line), p.b.x. etc. Routine measurements do not provide this information directly, because the trunks on which the traffic flow is recorded usually serve traffic from, and to, several classes. It can, however, usually be derived from traffic measurements on different exchanges.

Suppose there are k classes of line and N exchanges ($N \geqslant K$).

Let A_i = total traffic originated in exchange i

$n_{i,j}$ = number of lines in class j connected to exchange i

a_j = average traffic per line for class j

If all lines in class j had the same calling rate a_j, the following equations would hold exactly:

$$n_{11}a_1 + n_{12}a_2 + ... + n_{1k}a_k = A_1$$

$$n_{21}a_1 + n_{22}a_2 + ... + n_{2k}a_k = A_2$$
$$\text{etc. to}$$
$$n_{N1}a_1 + n_{N2}a_2 + ... + n_{Nk}a_k = A_N$$

In practice, because of variations in the traffic on different lines of the same class, these equations are unlikely to be satisfied exactly by any set of values of a_1, a_2 etc., unless $N = k$, when the a terms can of course be determined by solving the set of simultaneous equations. If possible, however, a larger sample of exchanges should be used. The number of equations could of course be reduced to k by the combination of exchanges, but better estimates can be obtained by the method of least squares. This consists in choosing the a_j values so as to minimise the sum of the squares of the differences between measured and calculated traffice for each exchange,[83] i.e.

$$S^2 = \sum_{i=1}^{N} \left(\sum_{j=1}^{k} n_{ij}a_j - A_i \right)^2$$

A necessary condition for a minimum is

$$\left(\frac{\partial S^2}{\partial a_j} \right) = 0$$

Hence

$$m_{11}a_1 + m_{12}a_2 + \dots + m_{1k}a_k = M_1$$
$$m_{21}a_1 + m_{22}a_2 + \dots + m_{2k}a_k = M_2$$
etc. to
$$m_{k1}a_1 + m_{k2}a_2 + \dots + m_{kk}a_k = M_k$$

where

$$m_{uv} = \sum_{i=1}^{N} n_{iu}n_{iv}$$

$$M_j = \sum_{i=1}^{N} n_{ij}A_i$$

We now have k equations to determine $a_1, a_2, \dots a_k$. If any of these values turns out to be zero or negative, this is an indication that the sample of exchanges is too small Exchanges should be chosen with widely different proportions of each class of lines; otherwise the sample may contain insufficient information to give a significant result.

Chapter 13

Traffic prediction

To predict the future growth of a network, it is necessary to predict the numbers of subscribers in different categories, such as business and residential, during the period concerned. Historical data should be examined for significant correlations between the demand for telephone service and other measures of economic activity, such as gross national product, income etc. By this means, forecasts of the overall growth of a national telephone network can be related to economic forecasts. Such forecasts are sometimes based on the assumption that the relative increment in demand for a service is proportional to the relative increment in economic activity.[124] This leads to a law of the form

$$\log Q = C_0 + C \log Y$$

where $Q =$ number of telephone subscribers at a certain time (or another suitable measure of demand for telephone service).

$Y =$ measure of economic activity

$C_0, C =$ empirical constants determined by analysis of historical data.

At the same time, forecasts of the number of subscribers in each area should be made by means of detailed market research. The results of the national and local forecasts may be inconsistent, in which case adjustments are necessary to arrive at an agreed basis for planning.

The average busy-hour traffic originated per line in each category is determined by traffic measurement (Section 12.9). This may require adjustment to allow for any foreseeable factors which may affect demand, such as changes in tariff policy, introduction of subscriber trunk dialling etc. Having determined the total originated traffic at each exchange, the next step is to distribute it to different destinations, including the same exchange. In this Section, we are concerned only with the prediction of *demand*. It will therefore be assumed that, in a self-contained area, the total originated and terminated traffic loads are equal, switching time and congestion loss being ignored. Ringing time, however, whether or not the call is answered, is included in the estimated traffic, since a call which receives ringing tone is effective so far as the switching system is concerned. The same applies to time spent in listening to busy tone when this is due to the called subscriber being engaged.

13.1 Coefficients of preference

One widely used method is to distribute the originated traffic at each exchange in proportion to the total traffic originated at the destination exchanges, multiplied by a 'coefficient of preference' which represents the degree of community of interest between the pair of exchanges concerned.

Let N = number of exchanges

O_i = total originated traffic at the ith exchange

k_{ij} = coefficient of preference for destination j by calls from exchange i

X_{ij} = traffic originating at i and terminating at j

Then

$$X_{ij} \equiv \frac{O_i O_j k_{ij}}{O_1 k_{i1} + O_2 k_{i2} + ... + O_N k_{iN}}$$

For convenience, k_{ii} may be taken as unity. In general,

$$k_{ij} \neq k_{ji}$$

It should be noted that the following variation of the formula is incorrect, because it does not automatically satisfy the obvious condition $\sum_j X_{ij} = O_i$ for all i.

$$X_{ij} = \frac{O_i O_j k_{ij}}{O_1 + O_2 + ... + O_N}$$

The coefficient of preference may be interpreted as follows:

Let $p(j|i)$ = conditional probability that an erlang, originating at i, terminates at j

$\phi(i|j)$ = conditional probability that an erlang, terminating at j, originates at i.

We have defined these probabilities in terms of erlangs rather than calls, to allow for differences in average holding time between calls to different destinations.

Then

$$X_{ij} = O_i p(j|i) = O_j \phi(i|j)$$

(assuming originated and terminating traffic are equal at each exchange)

Also

$$O_i = \sum_j X_{ij} = \sum_j O_j \phi(i|j)$$

thus

$$p(j|i) = \frac{O_j \phi(i|j)}{\sum_j O_j \phi(i|j)}$$

therefore

$$X_{ij} = \frac{O_i O_j \phi(i|j)}{\sum_j O_j \phi(i|j)}$$

Therefore
$$k_{ij} = \frac{\phi(i\,|\,j)}{\phi(i\,|\,i)}$$

If all the X_{ij} traffic flows are known, there are sufficient equations to determine the k_{ij} terms, taking k_{ii} as unity for all i. If future changes in the total originating at each exchange O_i can be predicted, the resulting changes in traffic distribution can be estimated without explicitly calculating the k_{ij} terms, assuming that relative community of interest throughout the network is unchanged, and that no new exchanges are added. Let O_i, O_j and X_{ij} denote the future values of traffic flows and let O_i^0, O_j^0 and X_{ij}^0 denote their present values. Then

$$k_{ij} = \frac{X_{ij}^0(O_1^0 k_{i1} + O_2^0 k_{i2} + \ldots + O_N^0 k_{iN}}{O_i^0 O_j^0}$$

Hence $\quad k_{ij} = (X_{ij}^0 O_i^0 k_{ii})/(X_{ii}^0 O_j^0) = (X_{ij}^0 O_i^0)/(X_{ii}^0 O_j^0)$

thus $\quad X_{ij} = (O_i O_j X_{ij}^0/O_j^0)/(O_1 X_{i1}^0/O_1^0 + O_2 X_{i2}^0/O_2^0 + \ldots + O_N X_{iN}^0/O_N^0)$

As a simple example, consider a network of three exchanges, the numbers of lines and traffic distribution being as shown in Table 14 below.

Table 14

		To exchange no.			Number of lines	Originating traffic	Average originating traffic per line
		1	2	3			
From	1	60 E	10 E	110 E	4 000	180 E	0·045 E
exchange	2	10 E	2 E	8 E	1 000	20 E	0·020 E
no.	3	110 E	8 E	700 E	10 000	818 E	0·0818 E
Terminating traffic		180 E	20 E	818 E			
Average terminating traffic per line		0·045 E	0·02 E	0·0818 E			

Suppose exchange number 1 is doubled in size, with no change in the calling rate, the other exchanges being unaltered. The calculations to find the traffic from exchange no. 1 to itself and to each of the other exchanges are set out in Table 15. This example, however, exhibits a weakness in the method. Calculating the new distribution of traffic from exchanges 2 and 3 in the same way, we obtain the following predicted traffic matrix (Table 16).

Table 15

Exchange no. (j)	Total originated traffic		Traffic from exchange no. 1 to each exchange		
	present (O_j^0)	future (O_j)	present (X_{ij}^0)	$\dfrac{O_j X_{ij}^0}{O_j^0}$	future (see note) (X_{ij})
1	180·0 E	360·0 E	60·0 E	120·0 E	180·0 E
2	20·0 E	20·0 E	10·0 E	10·0 E	15·0 E
3	818·0 E	818·0 E	110·0 E	110·0 E	165·0 E
			180·0 E	240·0 E	360·0 E

Note: The last column is calculated as follows:

$$X_{ij} = \frac{(\text{5th column}) \times O_1}{\text{total of 5th column}}$$

Table 16

		To exchange no.			Number of lines	Originating traffic	Average originating per line
		1	2	3			
From	1	180·0 E	15·0 E	165·0 E	8 000	360·0 E	0·045 E
exchange	2	13·4 E	1·3 E	5·3 E	1 000	20·0 E	0·020 E
no.	3	193·9 E	7·1 E	617·0 E	10 000	818·0 E	0·0818 E
Terminating traffic		387·3 E	23·4 E	787·3 E			
Average terminating traffic per line		0·0484 E	0·0234 E	0·0787 E			

Hence, an extension to exchange no. 1, with no alterations being made to the other exchanges, and no change in the average traffic originated by one line at any exchange, theoretically produces an 8% increase in the average traffic terminating on one line at exchange no. 1, a 17% increase at exchange no. 2, and a 4% decrease at no. 3. This redistribution of traffic, although not impossible, is unexpected; it raises a suspicion that the formula might produce completely unacceptable predictions in certain circumstances. For example, a small exchange in a large network might theoretically be swamped with traffic, possible more than 1 E per line in extreme cases, purely as a result of growth elsewhere. Such anomalies are conceivable with any prediction model which merely distributes the originating traffic without regard to the resulting distribution of terminating traffic.

13.2 Double-factor transformation

This anomaly suggests that it might be better to start with forecasts of both the total originating and the total terminating traffic at each exchange, and to distribute the total originating traffic in proportion to the terminating traffic. The resulting distribution may not be consistent with the assumed terminating traffic values, so an iterative procedure is necessary to reconcile the two. A suitable algorithm is the double-factor transformation devised by Kruithof,[59] which anticipated later work in connection with transportation theory; it seems to have been rediscovered independently by Furness[125] 28 years later. The method is as follows:

Let　X_{ij} = present traffic originating at i and terminating at j
O_i^0, O_1 = present and future traffic originating at i
T_j^0, T_j = present and future traffic terminating at j.

The present traffic distribution is represented by a matrix with elements X_{ij}, row totals O_{ij}^0, and column totals T_j^0. First, every row is multiplied by O_i/O^0, making the ith row total equal to O_i. Let T_j' denote the new total of the jth column. Each column is now multiplied by T_j/T_j' making the column totals equal to T_j, but altering the ith row total to O_i'. Each row is next multiplied by O_i/O_i' and the process is continued until the row and column totals agree with O_i and T_j, for all i and j, as closely as required. The convergence can be proved,[126] and, in practice, is usually rapid. The double-factor method is based on the assumption that the traffic originating at exchange i and terminating at exchange j is given by a formula of the type

$$X_{ij} = A_i B_j k_{ij}$$

k_{ij} is a measure of community of interest, which is assumed to remain constant during the prediction period. A_i and B_j are defined by the row and column totals, i.e.

$$\sum_j X_{ij} = A_i \sum_j B_j k_{ij} = O_i$$

$$\sum_i X_{ij} = B_j \sum_i A_i k_{ij} = T_j$$

Since A_i and B_j are, respectively, proportional to O_i and T_j, we have

$$X_{ij} \propto O_i T_j k_{ij}$$

which seems intuitively reasonable.

Let us apply this transformation to the previous example, assuming that the future terminated traffic at each exchange is the same as the originated traffic. The first row in the present traffic matrix is multiplied by 360/180, the second and third being unaltered, giving the following matrix; the future row and column totals are shown in brackets.

	1	2	3	Total
1	120·0 E	20·0 E	220·0 E	360·0 E (360·0 E)
2	10·0 E	2·0 E	8·0 E	20·0 E (20·0 E)
3	110·0 E	8·0 E	700·0 E	818·0 E (818·0 E)
Total	240·0 E	30·0 E	928·0 E	
	(360·0 E)	(20·0 E)	(818·0 E)	

The next step is to multiply the first, second and third columns by 360·0/240·0, 20·0/30·0, and 818·0/928·0, respectively, with the following results.

	1	2	3	Total
1	180·0 E	13·33 E	193·92 E	387·25 E (360·0 E)
2	15·0 E	1·333 E	7·052 E	23·385 E (20·0 E)
3	165·0 E	5·333 E	617·03 E	787·36 E (818·0 E)
	360·0 E	20·0 E	818·0 E	

The first, second and third rows are now multiplied by 360·0/387·25, 20·0/23·385 and 818/787·36, respectively, and so on. In short, the rows and columns are normalised alternatively until all the totals agree with the target values within acceptable accuracy. Close agreement is achieved in this example after only five iterations; the result is below.

	1	2	3	Total
1	170·41 E	12·95 E	176·64 E	360·00 E
2	12·96 E	1·18 E	5·86 E	20·00 E
3	176·63 E	5·87 E	635·50 E	818·00 E
	360·00 E	20·00 E	818·00 E	

The double-factor transformation can be proved to possess a number of properties which are desirable in any practical method of traffic prediction, including the following:

(*a*) *Uniqueness*
There is only one final matrix, having the specified row and column totals, obtainable from a given initial matrix by this method.

(*b*) *Reversibility*
The final matrix can be transformed into the initial matrix by the same procedure.

(*c*) *Transitivity*
The final matrix is the same whether it is derived from the initial matrix by a single transformation, or by way of a number of intermediate transformations.

(*d*) *Exchangeability*
If all the initial traffic flows are reversed the final flows will also be reversed.

(*e*) *Invariance under relabelling of exchanges*
Suppose, for example, that the first two rows are interchanged, then the first and

second columns. The double factor transformation is now applied, and the interchange of rows and columns is repeated. The final matrix is unaffected by the rearrangements.

(*f*) *Fractionability*
Another property which is ideally desirable is fractionability, i.e. exchanges may be combined or split without affecting traffic flows at other exchanges. The double-factor method does not meet this condition, but Kruithof states that the error is very small.

The traffic across the boundaries of the area under consideration can be allowed for by treating the outside world as equivalent to a single exchange, represented by an extra row and column in the matrix. If the external areas with which the one under consideration has significant community of interest are fairly well defined, it may be possible to make a rough estimate of the traffic circulating within that area (i.e. the 'own exchange' traffic of the hypothetical exchange), and thereby deduce the totals of the new row and column; the double-factor transformation can then be applied in the usual way. The following alternative approach suggested by Bear[62] was first employed in a computerised traffic-prediction model for system evaluation.[127] The figure at the intersection of the new row and column, i.e. the internal traffic of the hypothetical exchange is omitted. The transformation proceeds as before, except that the new row is excluded from row normalisation but its elements are included in column normalisation. Similarly, the new column is altered only by normalisation of the rows. If desired, more than one row and column can be used. It can be proved that this modified transformation is convergent and has the properties (*a*)–(*e*) listed above (G.T.J. Visick, private communication, 1972).

In practice, there may be zero elements in the traffic matrix, either because the traffic on certain routes is negligible, or because certain classes of traffic may be treated separately and deliberately excluded from the calculation. The double-factor method, in both its original and modified form, is generally applicable in such cases. If it fails to converge, the consistency of the future row and column totals should be checked. The transformation would not work, for example, in the following case, because the zero elements cannot be altered, and there are too few nonzero elements to produce the required totals.

Table 17

		Present matrix		Row totals	
				present	future
		1	0	1	3
		0	1	1	4
Column totals	present	1	1		
	future	1	6		

When the present matrix contains zero elements because some traffic flows are too small to be measured, it is generally advisable to substitute a small nominal value for zero to allow for realistic growth. The convergence of the matrix with zero elements should be checked first; otherwise, the substitution of nonzero elements might eliminate inconsistency, but the resulting matrix would be unreliable. Even if the original transformation converges, it is advisable to check that the quantities substituted for zeros are small enough not to have much effect on the rest of the transformed matrix.

A plausible justification for the double-factor transformation can be given in terms of information theory.[128] In the first place, let us divide each element in the matrix by the total originated (and therefore terminated) traffic in the network, so that the elements now represent the proportions of traffic between pairs of exchanges. Let us suppose for the moment that the future row and column totals are not constrained by forecasts of the total traffic at each exchange, and that we have no reason to foresee any changes in the relative community of interest. In that case, we would expect the proportion of traffic in each matrix element to remain constant, any other distribution would be surprising. If the row and column totals are constrained, it is in general, impossible to keep the traffic proportions constant, so the problem is to find the least surprising future distribution subject to these constraints.

In terms of information theory, the problem may be formulated as follows: If a matrix element changes from x_{ij} to y_{ij}, the information value of the change can be measured by $\log(y_{ij}/x_{ij})$, and the mean value of this gives the information value of the matrix transformation, i.e.

$$I(y:x) = \sum_i \sum_j y_{ij} \log(y_{ij}/x_{ij})$$

The 'least surprising' transformation is that which conveys the least amount of new information, and therefore minimises $I(y:x)$. Hence the required transformation is given by

$$\frac{\partial I(y:x)}{\partial y} = 0$$

subject to the new row and column totals

$$\sum_j y_{ij} = R_i$$

$$\sum_i y_{ij} = C_j$$

It can be shown by the method of Lagrange multipliers that the solution is of the form

$$y_{ij} = A_i B_j x_{ij}$$

13.3 Uniform community of interest

In some networks, such as those serving central urban areas, it may be fairly realistic to assume uniform community of interest; in other words, the probability of a randomly chosen subscriber calling another randomly chosen subscriber is independent of their locations, so that the whole network can be treated as a single exchange. In that case, the double-factor transformation is not required, because distribution of the originated traffic in proportion to the terminated traffic at each destination gives consistent results without iteration. The estimated traffic from i to j becomes

$$X_{ij} = \frac{O_i T_j}{T_1 + T_2 + ... + T_N}$$

thus

$$X_{1j} + X_{2j} + ... + X_{Nj} = \frac{T_j(O_1 + O_2 + ... + O_N)}{T_1 + T_2 + ... + T_N} = T_j$$

since the total originated and terminated effective traffic loads are equal.

Distribution in proportion to *originating* traffic. however, does not give correct terminating traffic values, unless the originating traffic at each exchange happens to be equal to the terminating traffic; for, if

$$X_{ij} = \frac{O_i O_j}{O_1 + O_2 + ... + O_N}$$

then

$$T_j = X_{1j} + X_{2j} + ... + X_{Nj} = O_j$$

Another method is to divide the traffic in proportion to the number of lines at the destination exchanges. This however automatically determines the ratio of originating to terminating traffic at each exchange, and may result in an unrealistic ratio when there are considerable differences in the average traffic per line at different exchanges.

13.4 Entropy and traffic distribution

A number of authors[37] have dealt with the application of the methods of statistical mechanics to the traffic distribution problem. If n is the total number of calls in a network during a certain period, and n_{ij} the number originating at i and terminating at j, then the number of ways in which calls might be allocated to routes, subject to a specified number of calls in each route, is

$$w = n! \left/ \left(\prod_{ij} n_{ij}! \right) \right.$$

The denominator is the product of all $n_{ij}!$ terms. It is argued[129] that the equilibrium traffic distribution corresponds to the most probable state. This is obtained by maximising w, or for convenience, $\log w$ (the entropy function), subject to specified row and column totals together with the condition that the total cost is fixed. The latter is

$$\sum_i \sum_j n_{ij} c_{ij}$$

where c_{ij} is the average cost of a call from i to j; this is not necessarily the actual monetary cost, but a 'generalised cost' related to the community-of-interest factor between i and j. The values of n_{ij} are determined by solving the set of differential equations

$$\frac{\partial w}{\partial n_{ij}} = 0$$

subject to
$$\sum_i n_{ij} = N_i$$

$$\sum_j n_{ij} = M_j$$

N_i and M_j being the total calls originating at i and terminating at j, respectively.

Also
$$\sum_i \sum_j n_{ij} c_{ij} = C, \quad \text{the fixed total cost}$$

The solution, by the method of Lagrange multipliers, is of the form

$$n_{ij} = A_i B_j e^{-\gamma c_{ij}}$$

Some variants of the entropy model lead to the same result without the assumption of fixed total cost.[37] Although similar in form, this result is not identical with that of the double-factor transformation. γ is a function of the c_{ij} terms and of the row and column totals,[130] so it is not invariant under change of the latter.

Although the entropy model leads to a community-of-interest function of negative exponential form, other forms are compatible with it.[131] The degree of deterrence associated with a generalised cost as perceived by the user, is not necessarily proportional to the value of c_{ij} as measured in monetary or equivalent units. For example, the effective cost as perceived by the user might be proportional to the logarithm of the measured value, in which case the community-of-interest function would be of the form

$$c_{ij}^{-\gamma}$$

The concept of entropy applies essentially to discrete entities like calls. It seems reasonable, however, to extend the results to traffic flow by analogy. Similar results have been obtained by purely econometric methods, without the use of this concept.[133] The entropy model has been successfully applied in transportation studies.[132] As a predictive method, it is less convenient for computation than the double-factor transformation. Since both methods lead to a law of the general form

$$X_{ij} \propto O_i T_j k_{ij}$$

they may be regarded as variants of the widely used gravity model. This term is derived by analogy with gravitational attraction, the parameters O_i and T_j being regarded as analogous with the masses of attracting bodies.

13.5 Other methods

A number of other methods of estimating traffic distribution have been suggested. Some of these do not automatically ensure consistency of interexchange traffic with total traffic, but discrepancies can easily be adjusted by means of the double-factor transformation. The following methods have been described by Rapp,[124] who remarks that 'they have at least the advantage of being based on clearly defined assumptions regarding the behaviour of the subscribers'. The first one is based on the assumption that the sum of the average traffic to the terminating exchange, produced by one subscriber at the originating exchange, and the average traffic from the originating exchange, received by one subscriber at the terminating exchange, is constant.

Thus, if A_{kl}^0, A_{kl} denote the present and future traffic from k to 1, and N_k^0, N_k the present and future numbers of subscribers connected to k and 1, we have

$$\frac{A_{kl}^0}{N_k^0} + \frac{A_{kl}^0}{N_l^0} = \frac{A_{kl}}{N_k} + \frac{A_{kl}}{N_l}$$

thus

$$A_{kl} = \frac{A_{kl}^0 \{ (N_k N_l / N_l^0) + (N_l N_k / N_k^0) \}}{N_k + N_l}$$

The second method minimises the sum of the squares of the change in the average traffic per subscriber, originated and received between a pair of exchanges. This implies $dF/dA = 0$

where

$$F = \{ (A_{k1}^0/N_k^0) - (A_{k1}/N_k) \}^2 + \{ (A_{k1}^0/N_l^0) - (A_{k1}/N_l) \}^2$$

from which

$$A_{kl} = \frac{A_{kl}^0 \{ (N_k^2 N_l / N_l^0) + (N_l^2 N_k / N_k^0) \}}{N_k^2 + N_l^2}$$

Another method, due to Ickler and Flachs, aims to minimise the change in subscribers' calling habits.[135] The conditional probability that a call, originating at i, terminates at j is

$$p(j|i) = X_{ij} \bigg/ \left(\sum_r X_{ir} \right)$$

The $p(j|i)$ terms can be arranged as a transition matrix with all row sums equal to unity. If $p(j|i)$ represents the present and $p'(j|i)$ the future probabilities, the $p'(j|i)$ terms are determined so as to minimise the perturbation in the matrix, as measured by the following expression:

$$\sum_i \sum_j \left(\frac{p'_{ij} - p_{ij}}{p_{ij}}\right)^2$$

The summation extends over all exchanges. This concept is somewhat similar to that of 'least surprise' which underlies the double-factor transformation.

13.6 Traffic distribution in new exchanges

When new exchanges are added to an existing network, they may be combined with appropriate neighbouring exchanges with comparable relative community of interest with the rest of the network. The matrix then has the same number of rows and columns as before, or possibly fewer if a new exchange is combined with more than one existing exchange. The traffic distribution after addition of new exchanges is derived from the previous distribution by a double-factor transformation. The new traffic elements are then divided as follows.

Suppose row/column no. 1 combines exchanges $i, j, k, \ldots$ while row/column no. 2 combines exchanges $I, J, K. \ldots$; some row/columns will probably have only one entry, being unaltered. The estimated traffic from I to j is

$$X_{Ij} = \frac{O_I T_j X_{21}}{(O_I + O_J + \ldots)(T_i + T_j + \ldots)}$$

where X_{21} = traffic from combined exchanges in row/column no. 2 to combined exchanges in row/column no. 1.

O_I = total traffic originated at I

T_j = total traffic terminated at j

Similarly
$$X_{IJ} = \frac{O_I T_J X_{11}}{(O_I + O_J + \ldots)(T_I + T_J + \ldots)}$$

The matrix is now enlarged to give every exchange a separate row/column, the calculated traffic values are inserted, and, if necessary, a double-factor transformation is applied to make the row and column totals agree with the predicted values of O_i, T_j etc. In general, traffic between existing exchanges will be altered by this transformation. This is reasonable, since the expansion is likely to effect the existing traffic distribution. It is, however, possible to keep particular routes unaltered if desired, by excluding them from the transformation and adjusting the row and column totals accordingly, provided sufficient variable elements are left to make this possible.

In estimating the totals of the rows and columns representing combined exchanges, allowance should be made for the fact that the subscribers may be transferred from existing exchanges to relieve congestion. An alternative method to the foregoing is to begin by calculating the initial traffic distribution, allowing for transferred subscribers only, as follows.

If a proportion ψ_{iK} of the subscribers served by an existing/exchange i are transferred to a new exchange K, a reasonable first estimate of the traffic from K to any existing exchange j, contributed by those subscribers, is $\psi_{iK}X_{ij}$, where X_{ij} is the traffic from i to j before the transfer. Similarly, the traffic to those subscribers from exchange j may be estimated as $\psi_{iK}X_{ji}$. The total traffic from K to j due to subscribers transferred to K is obtained by summation over all relevant exchanges i. Thus

$$X_{Kj} = \sum_i \psi_{iK} X_{ij}$$

and

$$X_{jK} = \sum_i \psi_{iK} X_{ji}$$

If there is more than one new exchange, the traffic between each one and the existing network is calculated as before, while the traffic between two new exchanges, K and L, is given by

$$X_{KL} = \sum_i \sum_j \psi_{iK} \psi_{jL} X_{ij}$$

$$X_{LK} = \sum_i \sum_j \psi_{iK} \psi_{jL} X_{ji}$$

the summation extending all exchanges contributing subscribers to K and L. If an exchange contributes to both K and L, $i = j$. The own-exchange traffic in i before transfer X_{ii} comprised $\psi_{iK}X_{ii}$ originated by the transferred subscribers, of which a proportion ψ_{iK} terminated on other transferred subscribers; hence, the own-exchange traffic in K due to transferred subscribers is

$$X_{KK} = \sum_i \psi_{iK}^2 X_{ii}$$

These first estimates of the traffic from and to transferred subscribers are inserted in the traffic matrix, the remaining traffic in the exchanges concerned being reduced correspondingly. A double-factor transformation is then applied to make the row and column totals agree with the estimated total originated and terminated traffic after the transfers. Finally, the effect of new subscribers is allowed for in the ordinary way.

13.7 Prior estimation of traffic distribution

When the foregoing methods are inapplicable, as in the case of a completely new network, the proportions in which the traffic from each exchange is divided between routes will have to be estimated in the light of the best local information available. This will probably be largely based on judgement and experience, but it may be possible to quantify the process to some extent in terms of coefficients of preference (Section 13.1). There may be ascertainable empirical relationships between the coefficients k_{ij} and various measurable parameters of the exchanges i and j, such as their distance apart d_{ij}. Laws of the form

$$k_{ij} = e^{-\gamma d_{ij}} \quad \text{or} \quad d_{ij}^{-\gamma}$$

(γ being an empirical constant) have been suggested. Again, the proportion of 'own-exchange' traffic might be related to such parameters as subscriber density, exchange size etc., enabling an estimate of X_{ii} to be made for a new exchange. The distribution of the rest of the traffic could then be calculated by the coefficient-of-preference method, omitting X_{ii}, from the empirical k_{ij} values. The suggested parameters are, of course, extremely crude measures of community of interest, but they may be useful for the purpose of deriving a plausible initial traffic distribution in the absence of better information, if there is reason to reject the hypothesis of uniform community of interest.

The resulting traffic matrix will probably be inconsistent with the estimates of total terminated traffic at each exchange, but the figures can be reconciled by means of a double-factor transformation. The same technique may be useful in long-range prediction, if the assumption of unchanging community of interest is considered implausible. Instead of deriving successive annual predictions from the initial distribution by repeated transformation, we may decide at some point to make a new prior estimate of traffic distribution, based on independent predictions of economic and other changes, and then apply the double-factor transformation to construct a consistent traffic pattern.

If a complete set of preference coefficients is known, the traffic distribution can be calculated directly by inserting them in the matrix and applying a double-factor transformation with the required row and column totals.

13.8 Prediction based on route traffic measurements

Ideally, the traffic flows on which predictions are based should be derived from direct measurements. With current techniques however, it is usually impracticable to collect routine measurements of traffic between each pair of originating and terminating exchanges, since the routes on which measurements are made may carry traffic from, and to, a number of origins and destinations. Most of the elements in the traffic matrix must therefore be based on estimates of relative community of interest between exchanges, possibly supplemented by special measurements involving counting of calls to different destinations at switching points. In either case, the estimated traffic flows may not be consistent with routine measurements on routes, and must be reconciled with them before using the data for prediction.[136]

So far, only effective traffic has been considered. The measured traffic, however, includes trunk occupations due to calls in the process of being set up, some of which may subsequently be lost, and these must be allowed for in the reconciliation process.[127] In the following simplified explanation, it is assumed that measurements have already been adjusted to eliminate ineffective traffic.

Before setting up the traffic matrix, it is necessary to check that the traffic flows are consistent.

Let O_a = originating effective traffic at exchange a

T_a = terminating effective traffic at exchange a

X_{aa} = effective 'own-exchange' traffic at exchange a

R_a = effective transit traffic switched via exchange a

M_a = measured effective traffic to other exchanges

m_a = measured effective traffic from other exchanges

Clearly, if the data are consistent, we have

$$M_a = R_a + (O_a - X_{aa})$$

$$T_a = (m_a - R_a) + X_{aa}$$

If all these traffic flows can be measured at the same time, these equations should be satisfied automatically, within the limits of switch-count error (Section 12.3); if not, the source of error should be sought. In some systems, however, it is not possible to measure all flows simultaneously, and the measured values may require adjustment to produce a consistent pattern.

We now have an effective traffic matrix, with row and column totals O_i and T_j based directly or indirectly on measurement. Some of the matrix elements X_{ij} may also be based on measurement, but most will probably be prior estimates adjusted to conform with the row and column totals. We also have a set of measurements of the traffic on each group of trunks. To reconcile the latter with the estimated matrix elements, we proceed as follows.

(i) Choose a pair of exchanges a and b with a direct route between them. Let M_{ab} denote the measured traffic on the route, and let L_{ab} denote the sum of all the X_{ij} terms contributing to the traffic carried on the route. Multiply each of these X_{ij} terms by M_{ab}/L_{ab}. Let $Y_{ij} = X_{ij}M_{ab}/L_{ab}$.

(ii) Repeat for all pairs of exchanges a and b with direct routes between them. Since an X_{ij} term may appear in more than one route (if it is switched via intermediate exchanges), it may give rise to more than one Y_{ij} term; in such cases, the mean value is used in subsequent calculation.

(iii) Apply a double-factor transformation to the Y_{ij} matrix to reconcile its elements with the row and column totals O_i and T_j. Let X'_{ij} denote the new value of X_{ij}, and let L'_{ab} denote the sum of all the X'_{ij} terms contributing to the traffic carried on route ab.

(iv) Repeat steps (i), (ii) and (iii) until M_{ab} and L'_{ab} agree within a specified degree of accuracy on all routes.

13.9 Exponential smoothing

While traffic forecasts should, of course, be based on present traffic measurements where possible, giving too much weight to these may tend to exaggerate the effect of temporary fluctuations. A simple method of using the results of measurement to update earlier forecasts, to obtain a smooth trend as a basis for current forecasts, is known as exponential smoothing.[136, 137]

Let M = measured traffic
A = previous forecast of M

The 'smoothed' value of the present traffic is

$$A_i = gM + (1-g)A$$

g is an empirical parameter between O and 1; a high value of g gives relatively high weight to current measurements as compared with past forecasts. The formula automatically gives more weight to recent forecasts than to earlier ones, and continually modifies them in the light of measurement. The choice of g is largely a matter of judgement and experience. It should obviously be nearly 1 if the interval since the previous forecast is too long to justify attaching much weight to it. If the interval between forecasts is very short, the data on which the previous forecast was based may be as significant as the current measurements, so a low value may be appropriate.

Chapter 14

Traffic simulation

Practical traffic problems are usually too complex for solution by theoretical analysis alone. The gap cannot, in practice, be filled by observation of real traffic, essential as this is for validating theoretical assumptions; field trials of experimental systems are expensive, and the impossibility of adequately controlling the test conditions makes the results of somewhat limited value to the traffic engineer. It has therefore been necessary to develop techniques for simulating traffic under controlled conditions. The main requirements for simulation are:

First, a model of the system, in sufficient detail to represent all states which are relevant to the investigation, and the transitions from one state to another.

Secondly, a means of generating the events which may alter the state of the system.

Thirdly, a set of rules describing how the system behaves in the case of any given event or combination of events.

In general, the events and the rules may be either probabilistic or deterministic. In telephone-traffic simulation, the events are usually probabilistic being dependent on subscribers behaviour, and the rules are usually deterministic, being dependent on the design of the system. Techniques for simulating random processes are often known as Monte Carlo methods.

Traffic-simulation tests were originally known as 'throwdown' tests or artificial traffic records. In early work, the instants of call origination were usually decided by dividing an arbitrary time period into, say, 10 000 short intervals, picking as many 4-digit random numbers as the required number of calls, and allocating the start of a call to each interval corresponding to a random number. A telephone directory was a convenient source of random numbers. Holding times were usually assumed to be constant, thereby automatically determining the end-points of calls. The model was usually a diagram on paper of the system to be simulated. The simulator worked through the calls in time order, marked the appropriate trunks busy in accordance with the hunting rules, and deleted the marks when the calls ended. At the same time, a record was kept of lost calls and any other necessary information, such as the number of calls in progress at regular intervals. The process was extremely laborious, since a prolonged simulation was necessary to obtain a statistically significant sample of lost calls at normal grades of service.

Later on, mechanical and electronic methods of generating random events were developed, to increase the speed of simulation. The latter were based on the conversion of a source of random noise into a stream of pulses. An electronic traffic analyser employed by the British Post Office[45] was a forerunner of the electronic random number indicator equipment (Ernie) used for drawing winning numbers of premium savings bonds. While producing much valuable information, these devices were insufficiently flexible to meet the great variety of simulation requirements arising from developments in switching systems. They have now been superseded by digital computers which are more flexible and easier to check for bias. They also permit exact replication of traffic for the comparison of different systems. which is not possible with devices depending on random noise.

Although modern computers make it possible to simulate very complex networks, this has by no means eliminated the need for theoretical analysis. The design and interpretation of simulations requires a sound theoretical basis. Moreover, once sufficient simulations have been carried out to validate an approximate theoretical formula, it is usually cheaper to use the formula for future calculations than to carry out further simulations. On the other hand, the simulation of a new system often leads to a better theoretical understanding of it, so that a combination of theory and simulation is more fruitful than either alone.

14.1 Random and pseudo-random numbers

While it would be possible to input a predetermined series of random numbers into a computer as data, it is more efficient to program the computer to generate the numbers mathematically. Since the result of any mathematical process is in principle completely predictable, numbers generated in this way are known as 'pseudo-random'. For practical purposes, however, they are indistinguishable from those produced by a physical random process, provided, they satisfy the usual statistical tests for randomness.

Thus, every digit must appear with equal frequency in the long run, and there must be no significant regularities. All pseudo-random number generating processes are cyclic, so it is essential that the cycle time be long enough to avoid repetition of identical patterns during one run.

There are a number of algorithms which meet these requirements,[140,141] one of which, the multiplicative congruential method, will be briefly described to illustrate the principle. An arbitrary number is chosen as a starting point, this is multiplied by an integer k, and the product is divided by an integer b, leaving a remainder R_1. The next term R_2, is the remainder when kR_1 is divided by b, and so on. The process is usually represented in the form

$$R_m = kR_{m-1} \pmod{b}$$

The pseudorandom numbers R_m can be interpreted either as integers or fractions according to the position of the decimal point. The numbers k and b are chosen to ensure an adequate cycle length while minimising the time of computation.[140]

Fig. 55 shows a common method of using random numbers to generate random intervals, such as intercall arrival times or holding times. $F(t)$ is the cumulative distribution function, i.e. the probability that the interval is less than t. A random number r is drawn from a uniform distribution in the range 0 to 1, and the interval t is calculated such that $F(t) = r$. If very many random numbers are drawn, the proportion less than any assigned value of $F(t)$ is obviously $F(t)$, so that the correct distribution of intervals is obtained.

While this method can be applied to purely empirical distributions, it is particularly suitable when $F(t)$ can be inverted in explicit form. An example is the negative exponential distribution, for which

$$F(t) = 1 - e^{-at}$$

$$\text{thus} \quad t = -\frac{\log_e \{1 - F(t)\}}{a}$$

$$= \frac{-\log_e (1 - r)}{a}$$

If the range of the distribution is finite and the probability density function is known, but the cumulative funciton cannot be expressed explicitly, the rejection method may be more suitable.

Let $f(t)$ be the p.d.f. of t; i.e. $f(t)dt$ is the probability of a value between t and $t + dt$. Let M be the maximum value of $f(t)$ and a and b the minimum and maximum values, respectively, of t. A random interval t is calculated from $t = a + (b - a)r$, where r is a random number from a uniform distribution between 0 and 1; then t is uniformly distributed between a and b. A second random number R is drawn in the same way and the value t is rejected if $R > f(t)/M$. The remaining values have the required distribution. This can be proved as follows.

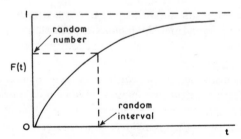

Fig. 55 Generation of random interval

The probability that a particular random interval lies between t and $t + dt$ is $dt(b - a)$, and the probability that it is accepted is $f(t)/M$. The prior probability of acceptance, before the value of the interval is known, is obtained by integrating the product of these probabilities over all values of t, giving

$$\int_a^b f(t)dt/\{(b - a)M\} = 1/\{(b - a)M\}$$

The conditional probability that an interval lies between t and $t + dt$, given that it is accepted, is therefore

$$f(t)dt/\{(b - a)M\} \div 1/\{(b - a)M\} = f(t)dt$$

14.2 System model

If a simulation is to be carried out using a digital computer, it is convenient to express the model of the system in the form of numbers and lists of numbers representing its structure and states. For example, the structure of the grading shown in Fig. 24*b* could be represented by a 4 x 10 array, in which each row represents a grading group, and columns 1, 2,...10 represent the corresponding outlet numbers. The element i, j contains a figure representing the number of the trunk connected to inlet j of grading group i. The traffic state of the grading at any time

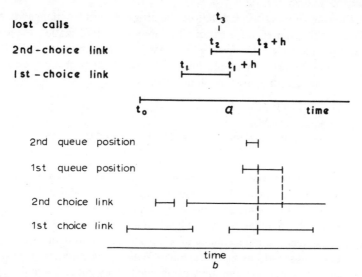

Fig. 56 Simulation of loss and delay systems

can be represented by a series of 20 digits '0' or '1', representing free and busy trunks, respectively; the nth digit corresponds to trunk number n. If a random number corresponding to a call arrival in group 1 is generated, the array is examined to determine the trunks available to that call, and the digit representing the state of the first free one is changed from 0 to 1. We are not always interested in the state of every separate trunk in the system. For example, it might be possible to save storage by representing only the total number of calls in a full-availability group.

14.3 Simulation of time and events

To simulate the passage of time with a digital computer, it must be treated as a discrete variable, the shortest interval being chosen so as to produce negligible error. To illustrate the principle, let us consider a 2-trunk full-availability group with Poisson input (A erlang), sequential search, constant holding time (h) and lost-calls cleared. A typical sequence of events is shown in Fig. 56a. Starting from an arbitrary time t_0 when there is no traffic, the first step is to generate the interval before the first call arrival. As the arrival process is Poissonian, this can be done by choosing a random interval t from a negative exponential distribution with mean value $1/A$. This gives the arrival time t_1 of the first call; $t_1 = t_0 + t$. Its termination time is, of course, $t_1 + h$. The arrival time of the second call is obtained by generating a second random interval starting at t_1, and so on. The third call, in this example, arrives when both trunks are busy and is lost. In general, each event may alter the state of the system and may generate future events. Thus, a call arrival, unless it is lost, increases the number of busy links by one, and generates a new call termination time, deterministically for constant holding times, and probabilistically in general.

Apart from performing the simulation, the program must be arranged to extract and print out the required information. The amount of information required will depend on the purpose of the investigation. Assuming that this is to verify Erlang's formulas for the distribution of the number of calls and probability of loss, the following information is required as a minimum:

(i) total number of calls offered
(ii) total number of calls lost
(iii) sample of observations of the number of calls in progress simultaneously.

The observations can be made at either regular or random intervals, so long as they are chosen without regard to the state of the system. It would not be correct to make observations that coincide with call endings, since the latter occur most frequently when the number of calls is high. Since the input is Poissonian in this instance, call arrivals are independent of the state of the system, so it would be correct to make observations at those times. With nonPoisson input, however, it would be better to determine the times of observation by random numbers. Several holding times should elapse before commencing observations, to establish equilibrium. The program should also be arranged to carry out the necessary calculations,

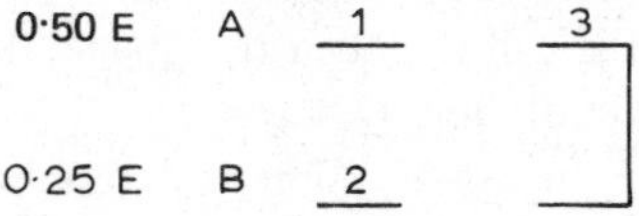

```
0·50 E    A    1       3
                          ⌐
                          |
0·25 E    B    2       ⌐__|
```

Fig. 57 Grading to illustrate roulette model simulation

and accuracy checks should be introduced. For example a goodness-of-fit test might be applied to check that the distribution of interarrival times is negative exponential. As the holding time of every call is known, the traffic offered can be calculated exactly.

Next, we will consider a simple delay system with two trunks, Poisson input variable holding time with a known distribution, first-in-first-out service (Fig. 56b) and a maximum capacity of two waiting calls. If we are interested in verifying a theoretical delay formula which assumes an infinite number of calls, which obviously cannot be simulated, the maximum queue capacity must be fixed so high that it is most unlikely to be reached.

As the holding time is variable, the end-points of the calls as well as their times of arrival are determined by generating random numbers. In the case of delayed calls, the end-points cannot be determined until the calls leave the queue. If we are interested in the distribution of delays, it is necessary to determine the time spent in the queue by each call (or possibly a sample of calls). If we are only interested in average delay, it is sufficient to know the average number of waiting calls, the total number of calls carried and delayed, and the duration of the simulated period.

14.4 Flow diagrams

A precise description of a simulation in ordinary language is usually extremely cumbersome; its essential logical structure can be more concisely displayed by means of a flow diagram. Fig. 58 shows a simplified flow diagram for the foregoing simulation of a 2-trunk delay system. The state of the system at any time and its changes are indicated by the values of the following variables:

A = time of next call arrival

T_x = time when trunk no. x is due for release ($x = 1$ or 2; $T_x = \infty$ if trunk x is already free)

L_x = 0 or 1 according to whether trunk x is free or busy

Q = number of waiting calls

C = number of calls completed since beginning of simulation

T = time of latest trunk seizure

M = total number of calls to be simulated..

Consider the system at the time of a call arrival; if L_1 = 0, it is changed to 1, indicating seizure of the first trunk. If $L_1 = 1$ and $L_2 = 0$, then L_2 is changed to 1. If both trunks are busy, the call is put in the queue; this is indicated by increasing Q by 1, or, in conventional notation, $Q = Q + 1$.

When a trunk is seized, its holding time is determined by generating a random interval H in accordance with the distribution law. This operation will be denoted 'sample H'. The release time of the trunk is obtained by adding H to the time of seizure and altering T_1 or T_2 accordingly; i.e.

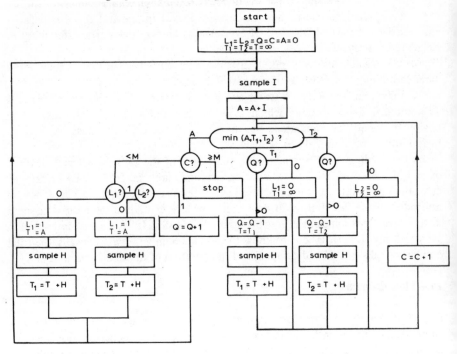

Fig. 58 Simulation flow diagram

$$T_x = T + H$$

where

$$T = \text{time of trunk seizure}$$

After a call has either seized a trunk or entered the queue, the time of the next call arrival is found by drawing a random interval I from a negative exponential distribution, since the input is Poissonian, and altering A accordingly

$$A = A + I$$

The time of the next event (arrival or termination) is now obtained by comparing the values of A, T_1 and T_2 and choosing the least, which is denoted $\min(A, T_1, T_2)$. This comparison is made after each event.

If the next event is a call arrival it is dealt with as before. If it is a termination on trunk x, and the queue is empty ($Q = 0$), the trunk release is indicated by putting $L_x = 0$. If $Q > 0$, L_x is not altered, but Q is reduced by 1 to indicate seizure of the trunk by a waiting call.

$$Q = Q - 1$$

The number of calls in the queue is counted at random intervals, say at times of call arrival or a sample thereof. The delay averaged over calls is given by

$$\frac{(\text{average value of } Q)(\text{total simulated period})}{\text{total number of calls simulated}}$$

The average delay on *delayed* calls is obtained by dividing the above expression by the proportion of calls entering the queue. The queue measurements and delay calculations are not shown in the flow diagram.

A detailed flow diagram contains all the information necessary to program the computer. While general-purpose languages are suitable for simulation programming, a number of special simulation languages have been developed to simplify programming. A simulation language may contain a single term for a common operation such as stacking calls in a queue for service in order of arrival, which is simple to perform in a manual simulation but has a complex logical structure. Some of these languages contain a greater range of facilities than are normally required in telecommunication-traffic simulation; while the number of calls and switching devices may be large, the number of different processes to be simulated is usually quite small. For this reason, simulation languages which are primarily designed for other applications with different requirements may be wasteful of memory capacity.

14.5 Roulette-model simulation

The foregoing method is known as time-true simulation. If the intercall arrival times and the holding times both have negative exponential distribution, the future of the system depends only on its present state, not on its previous history. A simpler technique, roulette-model simulation, can then be used.

Suppose we have k independent Poisson processes, with mean interarrival times $1/n_1, 1/n_2 \ldots 1/n_2 \ldots 1/n_k$. Since the interevent times in each process have negative exponential distribution, the probability that the next event in class 1 occurs between t and $t + dt$ after the present instant is $n_1 \exp(-n_1 t)dt$. The probability that no event in class 2 precedes this event is $\exp(-n_2 t)$, and the probability that no other event does so is

$$\exp(-n_2 t) \exp(-n_3 t) \ldots \exp(-n_k t) = \exp(-n_2 - n_3 - \ldots n_k)t$$

The probability that the next event in class 1 occurs between t and $t + dt$, *and* is not preceded by any other event, is therefore

$$n_1 \exp\{-(n_1 + n_2 + ... + n_k)t\}$$

The probability that the next event *is* in class 1 is obtained by integrating this expression from $t = 0$ to ∞, giving

$$n_1/(n_1 + n_2 + ... + n_k)$$

To illustrate the application of this result to simulation, consider a 2-group grading of three trunks at availability 2. Suppose the groups are unequally loaded, the traffic offered being 0·5 E and 0·25 E (Fig. 57). Both groups have the same average holding time. There are five classes of independent events; these are

(i) release of trunk no. 1
(ii) release of trunk no. 2
(iii) release of trunk no. 3
(iv) arrival of call in group A
(v) arrival of call in group B.

Suppose, in the first instance, that all three trunks are engaged. The expected time before the next event is 1 for classes I, II and III, 1/0·5 for class IV, and 1/0·25 for class V, in terms of the average holding time. Hence, the probability that the next event is in class IV, for example, is

$$0·5/(1 + 1 + 1 + 0·5 + 0·25)$$

and so on.

We must therefore associate each class of event with a range of random numbers such that the probabilities of the next event being in classes I, II, III, IV and V are in the ratio

$$1 : 1 : 1 : 0·5 : 0·25$$

One method is as follows; r denotes a random number from a uniform distribution.

If $0 \leqslant r < 1$, the next event is in class I
If $1 \leqslant r < 2$, the next event is in class II
If $2 \leqslant r < 3$, the next event is in class III
If $3 \leqslant r < 3·5$, the next event is in class IV
If $3·5 \leqslant r < 3·75$, the next event is in class V.

If a random number indicates an event in classes I, II or III, but the trunk in question is already free, the event is ignored, and another random number is generated. Since events are generated one at a time, there is no need for a list of future events. This results in a considerable saving in computation time and storage as compared with the time-true method.

Roulette-model simulation is applicable to finite-source traffic, provided the holding time has negative exponential distribution. If a is the calling rate per free source, and x out of N sources are busy, the expected time before the next call arrival is

$$1/\{(N-x)a\}$$

holding times. Thus each change of stage alters the relative probabilities of the next event being an arrival or termination, but it is still true that the next event depends only on the present state. With finite sources, the value of traffic offered can only be approximately determined in advance of simulation, since it depends on the loss.

In addition to the negative exponential distribution, the method can be applied to the Erlangian and hyperexponential distributions, which are derived from it (Section 7.6). If however, the number of stages is very large, time-true simulation may be preferable.

State probabilities can be determined in the roulette method by introducing random observation of the system as an additional class of random event; the probability of occurrence is determined so as to ensure the required frequency of observation, having regard to the accuracy of the results.

14.6 Simulation of repeated attempts

A realistic traffic simulation, particularly of a common-control system, should include repeated attempts (Section 9.1). The most frequent cause of these is usually the called line being already engaged. The most accurate way of allowing for this would be to include the state of every subscriber's line in the model, with a realistic distribution of line occupancy; this, however, would be expensive in storage requirements. Alternatively, every call could be given an average failure probability, based on observation if possible; otherwise, a rough estimate of the probability could be the average traffic per line in erlangs. The calls offered should be increased by a factor equal to the average number of attempts per call, which must be determined by observation. As an approximation the same failure probability applies to new and repeat calls. Failed calls contribute to the control traffic but not to the speech traffic. A still simpler method is to make no distinction between calls which find the called line free or busy, but simply to average the control and speech traffic over all calls, taking account of the extra traffic due to repeat attempts.

Given adequate provision of equipment, repeat attempts due to trunk congestion are few compared with those due to called lines being engaged, and can be allowed for by a slight estimated increase in the failure probability. If, however, the behaviour of the system under very heavy overloading is under investigation, a more realistic model may be desirable, including probability of calls being abandoned, repetition rate and number of calls awaiting repetition.

14.7 Accuracy of simulation

To assess the accuracy of estimated losses, delays etc, based on simulation, it is desirable to divide the run into a number (n) of intervals, and record the losses in each one. If the mean number of losses per interval is L, and its standard devia- is s, the 95% confidence range is very approximately

$$L \pm 2s/\sqrt{n}$$

This, of course, is based on the normal distribution; while the distribution of losses is usually far from normal, the error may not be serious when applied to the average of a large sample of intervals.

Kosten *et al.*[152] have shown that, for a full-availability-group loss system of N trunks, offered A erlang Poisson traffic, the coefficient of variation (ratio of standard deviation to mean) of the total losses during a given time is approximately

$$C_L = \{(N+A)/(N-A)AE_{1,N}(A)t\}^{\frac{1}{2}}$$

The coefficient of variation of the total congestion time during t is approximately

$$C_T = [2/\{E_{1,N}(A)t(N-A)\}]^{\frac{1}{2}}$$

Using exact expressions for C_L and C_T, it can be shown that $C_T < C_L$ so that, at least in the case of Poisson input, observation of congestion time gives a better estimate of congestion than observation of losses.

14.8 Cost-reduction techniques

Simulation is an expensive technique, and it is desirable to avoid using more memory capacity and run times than are necessary for the degree of accuracy required. Some of the techniques which can be used to improve the efficiency of simulation are described below.[144,140]

(a) Partial simulation
It is not generally essential to simulate the whole trunking scheme in detail. Fig. 3 will serve to illustrate the principle, although, in practice, such a simple scheme could easily be simulated completely. If, however, it were desired to simplify the simulation it might be sufficient to simulate one A switch and one C switch only, with their links to the B switches. It would be necessary to generate calls between these and other A and C switches, but only one end of the paths occupied by these calls would be simulated. The time congestion could be measured by observing the state of the links at random intervals, and recording the proportion of observations with no free pair of matching links. Care must be taken that the loss of detail is not such as to seriously distort the traffic pattern in the simulated portion.

As the examples of simulation already described illustrate, the seizure and release of trunks can be treated as simple point-events; it is not necessary to simulate circuit operations in detail; except in special cases, such as tests of double-connection risks.

(b) Control variables

In any finite run, the value of any parameter is likely to differ slightly from the true value for the population from which it is drawn. If the parameter to be measured is correlated with another parameter, the correlation can be used to derive a better estimate of the former than would be obtained by merely taking its average value.

Suppose, for example, that simulation is designed to measure the average delay in a waiting system with a specified traffic flow of A erlang, and that the observed delays and traffic values during a succession of equal runs are (w_1, A_1), (w_2, A_2) etc.

If w and A are the observed average delays and traffic flows for all runs, and A_0 is the nominal traffic flow generated in an infinite run, the corrected average delay is

$$w_0 = w + k(A_0 - A)$$

where k is the slope of the (A, w) curve, which can be determined by the method of least squares. A is known as a control variable. Similarly, if w depends on a number of control variables y_1, y_2 etc., the corrected value is

$$w_0 = w + \sum_i k_i y_i$$

(c) Antithetic variables

If each random number x_i (expressed as a probability) used to determine an event is replaced by its compliment $1 - x_i$, there will be negative correlation between sample values taken from the two runs. Thus, any parameter, such as loss probability which is above the true value in one run is likely to be below it in the other. The average of the two runs probably gives a better estimate than the average of two independent runs.

(d) Sequential analysis

The purpose of simulating a system is usually to determine how much traffic it will carry at a specified probability of congestion. It is necessary to make a preliminary estimate of the load to be simulated, based on some approximate formula. A number of runs with different loads may be required to obtain a range of congestion values sufficiently near the specified value for an accurate estimate of traffic capacity. The amount of simulation can be reduced by the use of sequential analysis. This technique, which is widely used in quality control, is a systematic procedure for drawing a statistical inference from a set of observations without the number of observations being predetermined; the procedure indicates when sufficient observations have been made to justify estimation of the unknown parameter within a given confidence range. Thus, the first simulation is terminated if the congestion so far observed indicates that the traffic is too far from the desired value, and a new simulation is started with an adjusted traffic value.

(e) Simulation of an approximate mathematical model

Grantges and Sinowitz[145] have described a computer program Neasim, which simulates, not the traffic system directly, but an approximate mathematical model of it, assuming Bernoulli distribution of link states. Even with this assumption theoretical calculation is often not feasible with practical networks, but it is relatively simple to simulate the paths available to the class of calls under consideration (Figs. 4b and 5b). Busy and idle states can then be assigned at random with the required probability for each link, according to its average occupancy. After each trial a check is made to determine whether or not there is a complete free path available; the proportion of trials with no free path gives an estimate of the blocking probability.

Appendix 1: Mathematical notes

An elementary knowledge of probability theory is essential to the understanding of the principles of telecommunication traffic. Some of the relevant material has been expounded at the appropriate points in the text. Other aspects which have not previously been explained are dealt with briefly in this Appendix; for a broader understanding of the subject, the reader is referred to standard textbooks, such as the works of Fry[52] or Feller.[146]

1 Combinatorial analysis

The number of ways in which r things can be arranged in a line is $r!$; for the first position can be filled by any one of r things; having filled the first position, the second can be filled in $r - 1$ ways, and so on.

The number of sets of r things which can be chosen out of n different things is

$$n!/\{(n-r)!\,r!\}$$

For the first thing can be chosen in n ways, the second in $n - 1$ ways, and so on. The total number of possible sequences of r choices is therefore

$$n(n-1)(n-2)\dots(n-r+1) = n!/(n-r)!$$

Any given set of r things, however, appears in $r!$ of these sequences by rearrangement. Hence the total number of sets of r things is $n!/\{(n-r)!\,r!\}$ which is usually denoted $\binom{n}{r}$ or ^nC_r. Similarly, the number of ways in which n things can be divided into k parcels, containing $r_1, r_2, \dots r_k$ things, is

$$n!/(r_1!\,r_2!\dots r_k!)$$

As an example, consider a group of four trunks, denoted a, b, c and d. Suppose they carry two calls at a given time. The number of different possible pairs of occupied trunks is $\binom{4}{2} = 6$, i.e. ab, ac, ad, bc, bd and cd. Note that the two calls are regarded as indistinguishable, since, as a rule, we are not interested in which calls occupies which trunk. If for some reason we do wish to distinguish the calls, by the calling parties' telephone numbers or in some other way, the number of ways in which they can be arranged is 12, since each different set of occupied trunks has two possible arrangements of calls.

2 Meaning of probability

The concept of probability, in its mathematical sense, applies primarily to possible events or states of affairs, where the circumstances are not known in sufficient detail to make their realisation or nonrealisation certain. Let us refer to a specific set of circumstances in which, or as a result of which, an event or state A may be realised, as a 'situation'; the realisation of A may or may not be contemporary with the situation. For practical purposes, the probability of A may be defined as the ratio of the number of situations in which A is realised to the total number of situations of the specified type.

In most applications, this ratio is regarded as the limiting value attained in a very large sample of situations. For example, if the situation consists in the offering of a telephone call to a specified group of trunks, and A denotes a lost call, the probability of A is the ratio of the number of lost calls to the total number of calls in a very large sample. In some cases, however, enough is known about the situation to enumerate all possible outcomes, and to state that these are all equally likely. The probability of A is then equal to the ratio of the number of outcomes in which A is realised to the total number; the definition in terms of frequency still applies, but the probability is known *a priori* without sampling. For example, if two calls are allocated at random to four trunks a, b, c and d, there are three arrangements, ab, bc, and cd, in which the calls are on adjacent trunks; the probability of this being the case is therefore 3/6. Probability values 1 and 0 are equivalent to certainty and impossibility, respectively.

If $A_1, A_2, ... A_r$ are mutually exclusive outcomes of a situation, with probabilities $p(A_1), p(A_2) ... p(A_r)$, the probability that one of them will occur is, by definition,

$$p(A_1) + p(A_2) + ... + p(A_r)$$

If there are no other possible outcomes, this expression is equal to one, since realisation of one of these outcomes is certain. Hence, if $p(A)$ and $p(\bar{A})$ denote the probabilities of realisation and nonrealisation of A, respectively, then

$$p(\bar{A}) = 1 - p(A)$$

Two probabilities, the sum of which is one, are known as complementary.

3 Conditional probability

$p(A \mid B)$ denotes the conditional probability of A, given the occurrence of another event B. It is defined, in the same way as the ordinary probability $p(A)$, as the ratio of the number of situations, $n(A, B)$, in which both A and B are realised, to the number, $n(B)$, in which B at least is realised. If n denotes the total number of relevant situations, we have

$$p(A \mid B) = n(AB)/n(B) = (n(AB)/n)(n/n(B)) = p(AB)/p(B)$$

where $p(AB)$ is the probability that A and B are both realised. It follows that

$$p(AB) = p(A \mid B)p(B) = p(B \mid A)p(A).$$

For example, let A denote the state 'a group of four trunks contains two calls only, which occupy adjacent trunks', and let B denote the state 'a group of four trunks contains two calls only, one of which occupies the left-hand trunk (a)'.

Then $n(AB) = 1$ (trunks a and b occupied)

$n(B) = 3$ (pairs of trunks ab, ac, or ad occupied)

$$p(A \mid B) = 1/3 = (1/6)/(3/6) = p(AB)/p(B)$$

If A and B are mutually exclusive,

$$p(A \mid B) = p(B \mid A) = 0$$

If A and B are independent

$$p(A \mid B) = p(A)$$

Thus $\qquad\qquad\qquad p(AB) = p(A)p(B)$

Similarly, for any number of independent events, we have

$$p(ABC \ldots) = p(A)p(B)p(C) \ldots$$

4 Bernoulli or Binomial distribution

Suppose a number of situations are similar in the sense that their possible outcomes and the probabilities of those outcomes are the same for all situations, but that the outcomes of one situation have no influence on those of any others; in other words, the situations or events are independent. Successive tosses of a coin, or simultaneous tosses of a number of coins, are a familiar example of this type of situation.

Consider a set of n situations, each having probability p of a particular outcome A. The probability that a particular subset, say the 1st, 3rd, 8th etc. comprising r situations out of n, will all have the same outcome A, while none of the remaining situations have this outcome, is $p^r(1 - p)^{n-r}$.

The total number of subsets of r out of n is $\binom{n}{r}$. Hence the probability that n situations, or trials, result in r outcomes A and $n - r$ other outcomes is

$$\binom{n}{r} p^r (1 - p)^{n-r}$$

This is known as the Bernoulli or Binominal distribution.

As a check, note that the total probability of all outcomes is

$$\sum_{r=0}^{n} \binom{n}{r} p^r (1 - p)^{n-r} = (p + 1 - p)^r = 1$$

as expected.

5 Probability-density function

In the case of quantities which have a continuous distribution, such as the holding time of a telephone call, the probability of a particular value is not in general a mathematically useful concept, and is replaced by the probability of a value within a short interval. If T is a continuous variable, and the probability that a value of T lies between t and $t + dt$, where dt is very small, is $p(t)dt$, then $p(t)$ is known as a probability-density function (p.d.f.). The usual laws of probability still apply, summation being replaced by integration. Thus

$$\int_{-\infty}^{\infty} p(t)\,dt = 1$$

6 Mean and variance

If a discrete random variable X takes value x with probability $p(x)$, its mean value or *expectation* is as

$$E(x) = \sum x p(x)$$

the summation extending over all possible values of x. In the case of a continuous variable the expectation is

$$E(x) = \int x p(x)\,dx$$

(This notation for expectation must not be confused with Erlang's congestion function.)

The most common measure of variability is the variance, which is defined as the square of the mean deviation of the variable from its mean value; squaring ensures that positive and negative deviations do not cancel out. The square root of the variance is known as the standard deviation, and the ratio of the latter to the mean is known as the coefficient of variation; the latter is useful for comparing the variability of quantities with widely different mean values. The variance can be expressed as follows:

Let
$$E(x) = \bar{x}$$

Then
$$\mathrm{var}(x) = E\{(x - \bar{x})^2\} = E(x^2) - 2\bar{x}E(x) + \bar{x}^2$$
$$= E(x^2) - \bar{x}^2$$

In other words, the variance is the difference between the square of the mean and the mean of the square.

If two independent variables have mean values $\bar{x}$ and $\bar{y}$, and variances var(x) and var(y), the variance of their sum is

$$\mathrm{var}(x + y) = E\{(x + y - \bar{x} - \bar{y})^2\}$$
$$= E\{(x - \bar{x})^2\} + E\{(y - \bar{y})^2\} - 2E(x - \bar{x})E(y - \bar{y})$$

But
$$E(x - \bar{x}) = E(x) - \bar{x} = 0 = E(y - \bar{y})$$

(In other words, positive and negative deviations from the mean cancel out.) Hence

$$\mathrm{var}(x + y) = \mathrm{var}(x) + \mathrm{var}(y)$$

Similarly, for any number of independent variables, the variance of the sum equals the sum of the variances.

7 Variance of the sum of a random number of variable quantities

As an example, consider the variance of the total duration of the calls occurring during a certain period, taking account of the variability of both the number of calls and the call-holding time. (Section 12.4)

Let n = number of calls during the given period
$\quad\quad t$ = call duration
$\quad\quad \phi_n$ = probability of n calls during the given period
$\quad\quad x$ = total duration of calls occurring during the given period
$\quad\quad p(x|n)$ = p.d.f. of x given that the number of calls is n
$\quad\quad V_x$ = variance of x
$\quad\quad V_t$ = variance of t
$\quad\quad V_n$ = variance of n
$\quad\quad V_{x|n}$ = variance of x, given that the number of calls is n
$\quad\quad \bar{n}, \bar{t}, \bar{x},$ = mean values of $n, t,$ and x
$\quad\quad \bar{x}_n$ = mean value of x, given that the number of calls is n

Then
$$\bar{x} = \bar{n}\bar{t}$$

$$V_{x|n} = nV_t$$

By definition $V_{x|n} = \int p(x|n)(x - n\bar{t})^2 dx$

$$V_x = \sum_n \int \phi_n p(x|n)(x - \bar{x})^2 dx$$

$$= \sum_n \int \phi_n p(x|n)(x - n\bar{t} + n\bar{t} - \bar{x})^2 dx$$

$$= \sum_n \phi_n V_{x|n} + \sum_n (n\bar{t} - \bar{x})^2 \phi_n \int p(x|n)\, dx$$

since $\sum_n \phi_n \int p(x|n)(x - n\bar{t})(n\bar{t} - \bar{x})\, dx$

$$= \sum_n \phi_n (n\bar{t} - \bar{x}) \int p(x|n)(x - n\bar{t})\, dx$$

$$= \sum_n \phi_n (n\bar{t} - \bar{x})(\bar{x}_n - n\bar{t})$$

and $x_n = n\bar{t}$

Thus $V_x = V_t \sum_n \phi_n n + \bar{t}^2 \sum_n (n - \bar{n})^2 \phi_n$

$$= V_t \bar{n} + \bar{t}^2 V_n$$

8 Erlang's loss formula with nonintegral trunks

It is sometimes necessary, for computational purposes, to postulate a nonintegral number of trunks, the equivalent trunk group in the equivalent random method of calculating blocking in gradings etc., is an example (Section 4.5). Again, the derivative of the number of trunks with respect to traffic, at constant grade of service, is useful in network optimisation (Section 11.4; also Akimaru.[134] It is therefore convenient to express Erlang's loss formula in the following alternative form:

$$E_{1,\,N}(A) = \frac{A^N e^{-A}}{\Gamma(N, A)}$$

where
$$\Gamma(N, A) = \int\limits_A^\infty x^N e^{-x} dx$$

The following limiting properties can be proved:

$$\lim_{A \to 0} E_{1,N}(A) = 0, \qquad \lim_{A \to \infty} E_{1,N}(A) = 1 \quad (0 \leqslant N < \infty)$$

$$\lim_{N \to 0} E_{1,N}(A) = 1, \qquad \lim_{N \to \infty} E_{1,N}(A) = 0 \quad (0 \leqslant A < \infty)$$

$$\lim_{N \to \infty} E_{1,N}(0) = 0, \qquad \lim_{A \to \infty} E_{1,\infty}(A) = 0$$

$$\lim_{A \to \infty} E_{1,0}(A) = 1, \qquad \lim_{N \to \infty} E_{1,N}(\infty) = 1$$

$$\lim_{N \to 0} E_{1,N}(0) = 0, \qquad \lim_{A \to 0} E_{1,0}(A) = 1$$

$$\lim_{N \to \infty} E_{1,N}(\infty) = 1, \qquad \lim_{A \to \infty} E_{1,\infty}(A) = 0$$

In other words, if there are no channels, the loss is 1 unless there is no traffic. If the number of channels is infinite, the loss is zero unless the traffic offered is infinite. If both are zero or both infinite, the loss may be either 1 or zero.

Appendix 2: Miscellaneous problems

1

A traffic load of one erlang is offered to a full-availability group of three trunks. The average call duration is 2 min.

(i) What is the average number of calls offered per hour?
(ii) What is the probability that no calls are offered during a specified period of 2 min?
(iii) What is the proportion of lost traffic?
(iv) If the trunks are always tested in the same order, how much traffic is carried by each trunk?

2

If a traffic load of 15 E is offered to a full-availability group of 25 trunks, the probability of congestion is 0·005. What is the probability that only one trunk is free? How often will a state of congestion occur during the busy hour, and how long will it last, on the average?
(Assuming the average holding time is 6 min.)

3

Using the traffic-capacity table in Appendix 3, determine the minimum number of trunks required to handle offered traffic loads of (a) 0·30 E, (b) 11·0 E, (c) 70·0 E at the following grade of service:

Loss not to exceed 0·005, with the proviso that the loss shall not exceed 0·01 in the event of either a 10% traffic overload, or a single trunk failure, not occurring at the same time.

4

Calculate the time congestion by Jacobaeus' method for the link system shown in Fig. 34 in the following cases:

(a) Bernoulli distribution in the B and C stages
(b) Bernoulli distribution in the B stage, Erlang distribution in C stage
(c) Erlang distribution in the B and C stages.

The data are as follows: $m = 10$; traffic in each B and C column is $3 \cdot 0$ E (the difference between offered and carried traffic may be neglected); in cases (a) and (b) there is no expansion or concentration at the A stage.

(d) Repeat the calculation for case (c) using the method of effective (average) availability, in conjunction with the original Palm–Jacobaeus formula. Assume that 10% of the traffic from the B column under consideration is destined for the C column under consideration.

(e) Repeat (d), but calculate the effective availability from case (a) instead of assuming it to be equal to the average availability.

The relevant values of $E_{1,N}(A)$ are as follows: If necessary, nonintegral values of N may be interpolated.

N	A	$E_{1,N}(A)$
2	3·0	0·529
3	3·0	0·346
4	3·0	0·206
5	3·0	0·110
10	3·0	0·00081
10	10·0	0·215

5

At a railway enquiry office dealing with telephone enquiries only, calls are queued in order of arrival until a clerk can deal with them. Suppose only one clerk is on duty, and the queue is designed to hold a maximum of 5 waiting calls. Calls arrive at the average rate of 60 per hour, and the average time taken to deal with an enquiry is 30 s, with negative exponential distribution. Assuming Poisson input, what is the probability that the queue is full? Calculate the waiting time in the queue, averaged over delayed calls and all calls.

6

A device which performs a certain switching function in a telephone exchange is required to commence operation within an average period of 10 ms after receiving a calling signal. It may be assumed that it responds to the signal instantaneously unless it is occupied by another call.

(a) If the device is held, on the average, for 50 ms per call, how many calls can it handle per hour?

(b) If the device is required to handle 18 000 calls per hour, what is the maximum value of the average holding time? Constant holding times and Poisson input may be assumed.

7

A full-availability group of 4 switches is observed at intervals of 1 min during 10 busy hours. Table 18 shows the number of occasions when various numbers of switches were observed to be engaged simultaneously.

Table 18

Number of switches engaged	Number of observations
0	89
1	164
2	173
3	114
4	60

The total number of lost calls was 34.

Assuming that time and call congestion are equal, estimate the average call duration and the average traffic offered, in erlangs, without using Erlang's formula.

8

Owing to a misprint in a newspaper advertisement, a residential line receives calls intended for a commercial firm, at the rate of 120 per hour. As the family is away, the calls are not answered.

(*a*) If an average caller listens to ringing tones for 30 s before clearing down, what is the probability of the line being engaged?

(*b*) If a caller who finds the line engaged clears down, on the average, after 3 s, what is the average number of callers receiving the line-engaged tone simultaneously?

Solutions to problems

1

(i) By definition, the traffic offered in erlangs is equal to the average number of calls offered per average holding time. The average number of calls offered per hour is therefore

$$1 \times 60/2 = 30$$

(ii) The average number of calls offered during a period of 2 min is 1. The probability that no calls are offered in this time, with Poisson input, is $e^{-1} = 0.368$

(iii) By Erlang's lost-call formula, the proportion of lost calls is 1/16.

(iv) The first, second and third choices carry 0·5 E, 0·3 E and 0·1375 E, respectively.

2

By Erlang's lost-call formula, the probability that 24 trunks are busy is

$$\frac{\dfrac{15^{24}}{24!}}{1 + \dfrac{15^1}{1!} + \dfrac{15^2}{2!} + ... + \dfrac{15^{25}}{25!}} = \frac{\dfrac{25}{15} \times \dfrac{15^{25}}{25!}}{1 + ... + \dfrac{15^{25}}{25!}}$$

$$= \frac{25}{15} \times 0\cdot005 = 0\cdot0083$$

Average number of calls per busy hour = $15 \times 60/6 = 150$. Average number of states of congestion (i.e. all trunks busy) during busy hour = average number of calls which find one trunk free

$$= 150 \times 0\cdot0083 = 1\cdot25$$

Average total duration of congestion states during busy hour = $3600 \times 0\cdot005 = 18$ s
Thus, average uninterrupted duration of a congestion state

$$= 18/1\cdot25 = 14\cdot4\,\text{s} = 360/25 \quad \text{as expected}$$

3
When the traffic lies between two tabulated values, trunks are read to the next higher whole number.

(a) $0\cdot30\,e$ requires 3 trunks at $0\cdot005$ loss.
    $0\cdot33\,e$ requires 3 trunks at $0\cdot01$ loss.
    $0\cdot30\,e$ requires 3 trunks at $0\cdot01$ loss, so 4 must be provided to allow for one out
        of order.
(b) 21
(c) 93

4
(a) Link occupancy = $3/10$ for A, B and C stages.

    Thus $E_L = \{2 \times 0\cdot3 - 0\cdot3^2\}^{10} = 0\cdot00119$

(b) $E_L = \dfrac{E_{1,10}(3\cdot0)}{E_{1,10}(3\cdot0/0\cdot30)} = \dfrac{0\cdot00081}{0\cdot215} = 0\cdot00377$

(c) $E_L = E_{1,10}(3\cdot0)(10 + 1 - 3\cdot0) = 0\cdot0065$

(d) Erlang distribution implies Poisson input, so time and call congestion are equal.
    Inlet blocking $B_L = E_{1,10}(3\cdot0) = 0\cdot00081$
    $\phi = 0\cdot10$

    Average availability of route C is

$$K = 10 \times 0\cdot30 \times 0\cdot10 + 10 \times 0\cdot7 = 7\cdot3$$

Thus, route blocking

$$B_R = \frac{E_{1,10}(3\cdot0)}{E_{1,2\cdot7}(3\cdot0)}$$

Interpolating to first differences only

$$E_{1,2\cdot7}(3\cdot0) = 0\cdot346 + 0\cdot183 \times 0\cdot3 = 0\cdot401$$

Thus

$$B_R = \frac{0\cdot00081}{0\cdot401} = 0\cdot0020$$

Total blocking $= 0\cdot0008 + 0\cdot0020 = 0\cdot0028$

(*e*) Effective availability K is given by

$$0\cdot3^K = 0\cdot00119 - 0\cdot3^{10} = 0\cdot00119$$

($0\cdot3^{10}$ is the inlet blocking with Bernoulli distribution at B-stage) whence $K = 5.6$

Route blocking

$$B_R = E_{1,10}(3\cdot0)/E_{1,4\cdot4}(3\cdot0)$$
$$= 0\cdot00081/0\cdot168 = 0\cdot0048$$

(interpolating as before)

Total blocking $= 0\cdot0008 + 0\cdot0048 = 0\cdot0056$

5

The average time for which the clerk is occupied is 30 min per hour. He can therefore be regarded as equivalent to a single trunk delay system offered 0·5 E. The probability of 5 or more waiting calls is

$$E_{2,1}(0\cdot5)(0\cdot5/1)^5 = 0\cdot0156$$

As explained in Section 7.7, this is an approximate formula which ignores the effect of lost calls on the distribution of waiting calls.

Waiting time average over delayed calls

$$= \frac{30}{1 - 0\cdot5} = 60\,\text{s}$$

Waiting time averaged over all calls

$$= 0\cdot5 \times 60 = 30\,\text{s}$$

6

(*a*) Let $a =$ traffic carried by the device, in erlangs.

The delay averaged over all calls is

$$\frac{a(50)}{(1-a)} \text{ millisecond}$$

Note that a negative exponential distribution has been assumed.

Thus, maximum tolerable value of a is given by

$$\frac{50a}{(1-a)} = 10$$

Therefore, $a = 0.167\,\text{E}$

This represents

$$\frac{3600 \times 0.167}{0.05} = 12\,000 \text{ calls per hour.}$$

(*b*) Average delay over all calls = $10\,\text{ms}$

$$= at/\{2(1-a)\}$$

where t is the average holding time in milliseconds.
But $a = Ct$ where C = average number of calls per millisecond.

$$= 18000/(60 \times 60 \times 10^3)$$

Thus

$$10 = \{(Ct)t\}/\{2(1-Ct)\}$$
$$= t^2/\{2(200-t)\}$$

This quadratic equation has only one positive solution, i.e. $t = 54\,\text{ms}$

7

Total number of observations = 600.
Probability of all trunks being busy (time congestion)

$$= 60/600 = 0.1$$

Estimated total number of calls offered = $34/0.1 = 340$.
Estimated number of carried calls = $340 - 34 = 306$.
But the average traffic carried is equal to the average number of simultaneous calls, which is

$$\frac{0 \times 89 + 1 \times 164 + 2 \times 173 + 3 \times 114 + 4 \times 60}{600} = 1.82 \text{ erlang}$$

Let h = average call duration in minutes

Then $\dfrac{\text{total calls} \times \text{average duration}}{\text{total time}} = \dfrac{306h}{10 \times 60} = 1.82$

Thus $h = 3.57\,\text{min}$

Traffic offered = $340 \times \dfrac{3.57}{600} = 2.02\,\text{E}$

8

Traffic offered to the line is

$$a = \frac{120 \times 30}{3600} = 1\,\text{E}$$

Probability of the line being engaged $= a/(1 + a) = 0\cdot5$

Average number of calls per hour encountering 'line engaged'

$$= 120 \times 0\cdot5 = 60$$

Average number of callers receiving line engaged tone simultaneously $= 60 \times 3/3600$
$= 0\cdot05$.

Appendix 3: Traffic-capacity Table for full-availability groups

Number of trunks	50 (0.02)	1 lost call in 100 (0·01)	200 (0·005)	1000 (0·001)	Number of trunks	50 (0·02)	1 lost call in 100 (0·01)	200 (0·005)	1000 (0·001)
	E	E	E	E		E	E	E	E
1	0·020	0·010	0·005	0·001	24	16·6	15·3	14·2	12·2
2	0·22	0·15	0·105	0·046	25	17·5	16·1	15·0	13·0
3	0·60	0·45	0·35	0·19	26	18·4	16·9	15·8	13·7
4	1·1	0·9	0·7	0·44	27	19·3	17·7	16·6	14·4
5	1·7	1·4	1·1	0·8	28	20·2	18·6	17·4	15·2
6	2·3	1·9	1·6	1·1	29	21·1	19·5	18·2	15·9
7	2·9	2·5	2·2	1·6	30	22·0	20·4	19·0	16·7
8	3·6	3·2	2·7	2·1	31	22·9	21·2	19·8	17·4
9	4·3	3·8	3·3	2·6	32	23·8	22·1	20·6	18·2
10	5·1	4·5	4·0	3·1	33	24·7	23·0	21·4	18·9
11	5·8	5·2	4·6	3·6	34	25·6	23·8	22·3	19·7
12	6·6	5·9	5·3	4·2	35	26·5	24·6	23·1	20·5
13	7·4	6·6	6·0	4·8	36	27·4	25·5	23·9	21·3
14	8·2	7·4	6·6	5·4	37	28·3	26·4	24·8	22·1
15	9·0	8·1	7·4	6·1	38	29·3	27·3	25·6	22·9
16	9·8	8·9	8·1	6·7	39	30·1	28·2	26·5	23·7
17	10.7	9·6	8·8	7·4	40	31·0	29·0	27·3	24·5
18	11·5	10·4	9·6	8·0	41	32·0	29·9	28·2	25·3
19	12·3	11·2	10·3	8·7	42	32·9	30·8	29·0	26·1
20	13·2	12·0	11·1	9·4	43	33·8	31·7	29·9	26·9
21	14·0	12·8	11·9	10·1	44	34·7	32·6	30·8	27·7
22	14·9	13·7	12·6	10·8	45	35·6	33·4	31·6	28·5
23	15·7	14·5	13·4	11·5	46	36·6	34·3	32·5	29·3

Number of trunks	1 lost call in				Number of trunks	1 lost call in			
	50 (0·02)	100 (0·01)	200 (0·005)	1000 (0·001)		50 (0·02)	100 (0·01)	200 (0·005)	1000 (0·001)
	E	E	E	E		E	E	E	E
47	37·5	35·2	33·3	30·1	74	62·9	59·8	57·3	52·6
48	38·4	36·1	34·2	30·9	75	63·9	60·7	58·2	53·5
49	39·4	37·0	35·1	31·7	76	64·8	61·7	59·1	54·3
50	40·3	37·9	35·9	32·5	77	65·8	62·6	60·0	55·2
51	41·2	38·8	36·8	33·4	78	66·7	63·6	60·9	56·1
52	42·1	39·7	37·6	34·2	79	67·7	64·5	61·8	56·9
53	43·1	40·6	38·5	35·0	80	68·6	65·4	62·7	57·8
54	44·0	41·5	39·4	35·8	81	69·6	66·3	63·6	60·3
55	45·0	42·4	40·3	36·7	82	70·5	67·2	64·5	59·5
56	45·9	43·3	41·2	37·5	83	71·5	68·1	65·4	60·4
57	46·9	44·2	42·1	38·3	84	72·4	69·1	66·3	61·3
58	47·8	45·1	43·0	39·1	85	73·4	70·1	67·2	62·1
59	48·7	46·0	43·9	40·0	86	74·4	71·0	68·1	63·0
60	49·7	46·9	44·7	40·8	87	75·4	71·9	69·0	63·9
61	50·6	47·9	45·6	41·6	88	76·3	72·8	69·9	64·8
62	51·6	48·8	46·5	42·5	89	77·2	73·7	70·8	65·6
63	52·5	49·7	47·4	43·4	90	78·2	74·7	71·8	66·6
64	53·4	50·6	48·3	44·1	91	79·2	75·6	72·7	67·4
65	54·4	51·5	49·2	45·0	92	80·1	76·6	73·6	68·3
66	55·3	52·4	50·1	45·8	93	81·0	77·5	74·3	69·1
67	56·3	53·3	51·0	46·6	94	81·9	78·4	75·4	70·0
68	57·2	54·2	51·9	47·5	95	82·9	79·3	76·3	70·9
69	58·2	55·1	52·8	48·3	96	83·8	80·3	77·2	71·8
70	59·1	56·0	53·7	49·2	97	84·8	81·2	78·2	72·6
71	60·1	57·0	54·6	50·1	98	85·7	82·2	79·1	73·5
72	61·0	58·0	55·5	50·9	99	86·7	83·2	80·0	74·4
73	62·0	58·9	56·4	51·8	100	87·6	84·0	80·9	75·3

References

1　JENSEN, A.: 'Moe's principle' (KTAS. Copenhagen, 1950)
2　GRINSTED, W.H.: 'A study of telephone traffic problems with the aid of the principles of probability', *Post Off. Electr. Eng. J.,* 1915, **8**, pp. 33–45
3　ERLANG, A.K.: 'Solution of some problems in the theory of probabilities of significance in automatic telephone exchanges', *ibid.,* 1918, **10**, pp. 189–197
4　LEIGHTON, A.G., and KIRKBY, W.: 'An improved method of grading: the partially skipped grading', *ibid.,* 1972, **65**, pp. 165–172
5　BRETSCHNEIDER, G.: 'Exact loss calculations of gradings', preprints of technical papers, 5th International Teletraffic Congress, New York, 1967, pp. 162–169
6　O'DELL, G.F.: 'An outline of the trunking aspect of automatic telephony', *J. IEE,* 1927, **65**, pp. 185–222
7　WILKINSON, R.I.: 'Theories for toll traffic engineering in the USA', *Bell Syst. Tech. J.,* 1956, **35**, pp. 421–514
8　JACOBAEUS, C.: 'A study on congestion in link systems', *Ericsson Tech.,* 1950, **48**, pp. 1–70
9　LOTZE, A., and WAGNER, W.: 'Table of the modified Palm–Jacobaeus loss formula'. Institut fur Nachrichtenvermittlung und Datenverarbeitung der Technischen Hochschule, Stuttgart, 1963
10　MCHENRY, C.: 'An investigation of the loss involved in trunking from primary line switches to 1st selectors via secondary line switches in Strowger automatic exchanges', *Post Off. Electr. Eng. J.,* 1922, **14**, pp. 217–227
11　DUMJOHN, F.P., and MARTIN, N.H.: 'Experimental determination of traffic loads and congestion on first selectors using 10-contact first and second preselectors', *ibid.,* 1922, **15**, pp. 133–145
12　BININDA, N., and WENDT, W.: 'Die effektive Erreichbarkeit fur abnehmerbundelhinter Zwischenleitungsanordnungen', *Nachrichtentech. Z.,* 1959, **12**, pp. 579–585 and 1961, **14**, p. 40
13　KHARKEVICH, D.: 'An approximate method for calculating the number of junctions in a crossbar system exchange', *Elektrosvyaz,* 1959, **2**, pp. 55–63
14　ELLDIN, A.: 'On the dependence between the two stages in a link system', *Ericsson Tech.,* 1961, **2**, pp. 185–259
15　BININDA, N., and DAISENBERGER, G.: 'Recursive and iterative formulae for the calculation of losses in link systems of any description, preprints of technical papers, 5th International Teletraffic Congress, New York, 1967, pp. 318–326
16　BENES, V.E.: 'Traffic in connecting networks when existing calls are rearranged', *Bell Syst. Tech. J.,* 1970, **49**, pp. 1471–1482
17　BENES, V.E.: 'Mathematical theory of connecting networks and telephone traffic' (Academic Press, 1965)
18　COHEN, J.W.: 'On the fundamental problem of telephone traffic theory and the influence

of repeated calls', *Philips Telecommun. Rev.*, 1957, **18**, pp. 49–100

19 LE GALL, P.: 'Sur le taux d'efficacité et la stationarité du traffic téléphonique', *Commutat. & Electron.* 1971, **35**, pp. 7–36

20 DESCLOUX, A.: 'Delay tables for finite and infinite-source systems' (McGraw-Hill, 1962)

21 CROMMELIN, C.D.: 'Delay probability formulae', *Post Off. Electr. Eng. J.*, 1934, **26**, pp. 266–274

22 RIORDAN, J.: 'Stochastic service systems' (Wiley, 1962)

23 WILKINSON, R.I.: 'Working curves for delayed exponential calls served in random order', *Bell Syst. Tech. J.*, 1953, **32**, pp. 360–383

24 POVEY, J.A., and COLE, A.C.: 'The use of electronic digital computers for telephone traffic engineering', *Post Off. Electr. Eng. J.*, 1965, **58**, pp. 203–209

25 BURKE, P.J.: 'Equilibrium delay distribution for one channel with constant holding time, Poisson input and random service', *Bell Syst. Tech. J.*, 1959, **38**, pp. 1021–1031

26 ENGSET, T.: 'Die Wahrscheinlichkeitsrechnung zur Bestimmung der Wahleranzahl in automatischen Fernsprechamtern', *Electrotech. Z.*, 1918, **31**, pp. 304–305

27 MOLINA, E.C.: 'The theory of probabilities applied to telephone trunking problems', *Bell Syst. Tech. J.*, 1922, **1**, pp. 69–81

28 PALM, C.: 'Étude des délais d'attente', *Ericsson Tech.*, 1937, **5**, pp. 39–56

29 WILKINSON, R.I.: Discussion contribution on 'Basic theory underlying Bell System facilities capacity tables', by A.L. Gracey, *AIEE Trans.*, 1950, pp. 238–244

30 CLOS, C.: 'A study of nonblocking switching networks', *Bell Syst. Tech. J.*, 1953, **32**, pp. 406–424

31 KAPPELL, J.G.: 'Non-blocking and nearly non-blocking multi-stage switching arrays', preprints of technical papers, 5th International Teletraffic Congress, New York, 1967, pp. 238–241

32 DUERDOTH, W.T., and SEYMOUR, C.A.: 'A quasi-non-blocking TDM switch', Proceedings of the 7th International Teletraffic Congress, Stockholm, 1973, pp. 632/1–4

33 KOSTEN, L.: 'Application of artificial traffic methods to telephone problems', *Teleteknik (Engl. ed.)*, 1957, **1**, pp. 107–110

34 GOLEWORTH, H.M.G., KYME, R.C., and ROWE, J.A.T.: 'The measurement of telephone traffic', *Post Off. Electr. Eng. J.*, 1972, **64**, pp. 227–233

35 HAYWARD, W.S., Jun.: 'The reliability of telephone traffic load measurements by switch counts', *Bell Syst. Tech. J.*, 1952, **31**, pp. 357–377

36 POVEY, J.A.: 'A study of traffic variations and a comparison of post-selected and time-consistent measurements of traffic', preprints of technical papers, 5th International Teletraffic Congress, New York, 1967, pp. 1–6

37 TOMLIN, J.A., and TOMLIN, S.G.: 'Traffic distribution and entropy', *Nature*, 1968, **220**, pp. 974–976

38 BROCKMEYER, E., HALSTROM, H.L., and JENSEN, A.: 'The life and works of A.K. Erlang' (Copenhagen Telephone Company, 1948)

39 DE BOER, J.: Comparison of random selection and selection with fixed starting position in a multi-stage link network', *Philips Telecommun. Rev.*, 1973, **31**, pp. 148–155

40 KUHN, P.: 'Waiting time distributions in multi-queue delay systems with gradings', Proceedings of the 7th International Teletraffic Congress, Stockholm, 1973, pp. 242/1–9

41 HIEBER, L.J.: 'About multi-stage link systems with queuing', 6th International Teletraffic Congress, Munich, 1970, pp. 233/1–7

42 CASEY, J., Jun., and SHIMASAKI, N.: 'Optimal dimensioning of a satellite network using alternate routing concepts', 6th International Teletraffic Congress, Munich, 1970, pp. 344/1–8

43 GIMPELSON, L.A.: 'Network management: design and control of communications networks', *Electr. Commun.*, 1974, **49**, pp. 4–22

44 KYME, R.C.: 'A system for telephone traffic measurement and routing analysis by computer', Proceedings of the 7th International Teletraffic Congress, Stockholm, 1973, pp. 525/1–6

45 BROADHURST, S.W., and HARMSTON, A.T.: 'Studies of telephone traffic with the aid of a machine', *Proc. IEE*, 1953, **100**, Pt. 1, pp. 259–274

46 Report of the 1st International Congress on the Application of the Theory of Probability in Telephone Engineering and Administration, Copenhagen, 1955, *Teleteknik, (Engl. edn.)*, 1957, **1**, pp. 1–130

47 Report of the 2nd International Congress on the Application of the Theory of Probability in Telephone Engineering and Administration, The Hague, 1958, *PTT-Bedr.*, 1960, 9, pp. 159–209

48 Report on the 3rd International Teletraffic Congress, Paris, 1961, *Ann. Telecommun.*, 1962, 17, pp. 145–226

49 Report on the Proceedings of the 4th International Teletraffic Congress, London, 1964, *Post. Off. Telecommun. J.*, special issue, pp. 1–66

50 BERKELEY, G.S.: 'Traffic and trunking principles in automatic telephony' (Ernest Benn, 2nd revised edn., 1949)

51 SYSKI, R.: 'Introduction to congestion theory in telephone systems' (Oliver & Boyd, 1960). Includes a chapter on telephone systems by N.H.G. Morris

52 FRY, T.C.: 'Probability and its engineering uses' (Van Nostrand, 2nd edn, 1965)

53 BECKMANN, P.: 'An introduction to elementary queuing theory and telephone traffic' (The Golem Press, Boulder, Colo., 1968)

54 ATKINSON, J.: 'Telephony' (Pitman, 1957)

55 RUBIN, M., and HALLER, C.E.: 'Communication switching systems' (Reinhold, 1966)

56 COX, D.R., and SMITH, W.L.: 'Queues' (Methuen, 1961)

57 HAYWARD, W.S., Jun. and WILKINSON, R.I.: 'Human factors in telephone systems and their influence on traffic theory, especially with regard to future facilities', 6th International Teletraffic Congress, Munich, 1970, pp. 431/1–10

58 WILKINSON, R.I.: 'Some comparisons of load and loss data with current teletraffic theory', *Bell Syst. Tech. J.*, 1971, 50, pp. 2808–2834

59 KRUITHOF, J.: 'Telefoonverkeersrekening', *De Ingenieur*, 1937, 52, pp. E15–E25

60 OLSSON, K.M., ANDERBERG, M., and LIND, G.: Report on the 7th International Teletraffic Congress in Stockholm, June 13–20, 1973, *Ericsson Tech.*, 1973, 29, pp. 107–144

61 BUCHNER, M.M., Jun., and NEAL, S.R.: 'Inherent load balancing in step-by-step switching systems', *Bell Syst. Tech. J.*, 1971, 50, pp. 135–165

62 BEAR, D.: 'Some theories of telephone traffic distribution: a critical survey', Proceedings of the 7th International Teletraffic Congress, Stockholm, 1973, pp. 531/1–5

63 KUMMERLE, K.: 'An analysis of loss approximation for link systems', preprints of technical papers, 5th International Teletraffic Congress, New York, 1967, pp. 327–336

64 LOTZE, A.: 'History and development of grading theory', preprints of technical papers 5th International Teletraffic Congress, New York, 1967, pp. 148–161

65 BRIDGFORD, J.N.: 'The geometric group concept and its application to the dimensioning of link access systems', Proceedings of the 4th International Teletraffic Congress, London, 1964, paper 13

66 BOTSCH, D.: 'International Standardising of loss formulae?', 5th International Teletraffic Congress, New York, 1967, pp. 90–95

67 DE FERRA, P., and MASSETTI, G.: 'On the quality of telephone service in an automatic network and its economical expression'. Proceedings of the 6th International Teletraffic Congress, Munich, 1970, pp. 135/1–6

68 KOSTEN, L.: 'On the validity of the Erlang and Engset loss formulae', *PTT-Bedr*, 1948, 2, pp. 42–45

69 MARTIN, N.H.: 'A note on the theory of probability applied to telephone traffic problems', *Post. Off. Electr. Eng. J.*, 1923, 16, pp. 237–241

70 RUBAS, J.: 'Analysis of congestion in small PABX's'. Proceedings of the 6th International Teletraffic Congress, Munich, 1970, pp. 211/1–8

71 HERZOG, U.: 'Calculation of two-way trunk arrangements with different types of traffic input', *ibid.*, 1970, pp. 217/1–6

72 BAZLEN, D.: 'Link systems with bothway connections and outgoing finite source traffic'. Proceedings of the 7th International Teletraffic Congress, Stockholm, 1973, pp. 316/1–8

73 RONNBLOM, W.: 'Traffic loss of a circuit group consisting of bothway circuits which is accessible for the internal and external traffic of a subscriber group', *Telete.*, 1959, no. 2 (English edition), pp.79–92

74 MARROWS, B.: 'Circuit provision for small quantities of traffic,' *Telecommun. J. Aust.*, 1959, 11, pp. 208–211

75 PARVIALA, A.: 'Calculation of the optimum number of trunk lines based on Moe's principle'. Proceedings of the 7th International Teletraffic Congress, Stockholm, 1973, pp. 422/1–5

76 LONGLEY, H.A.: 'The efficiency of gradings,' *Post Off. Electr. Eng. J.*, 1948, **41**, pp. 45–49 and 67–72

77 EINARSSON, K-A., HAKANSSON, L., LINDGREN, E., and TANGE, I.: 'Simplified, type of gradings with skipping', *Teleteknik (Engl. edn.)*, 1961, pp. 74–96

78 KRUITHOF, J.: 'Loss formulas for homogeneous gradings of the second order in telephone switching employing random hunting', *Electr. Commun.*, 1958, **35**, pp. 57–68

79 WILKINSON, R.I.: 'The interconnection of telephone systems – graded multiples', *Bell Syst. Tech. J.*, 1931, **10**, 531–564

80 BRETSCHNEIDER, G.: 'Die Berechnung von Leitungsgrupprn fur uberfließ enden Verkehr in Fersprechwahlanlagen', *Nachrichtentech. Z.*, 1956, **11**, pp. 533–540 (reprinted with additional reference data by Siemens & Halske Aktiengesellshaft)

81 COHEN, J.W., and BEUKELMAN, B.J.: 'Call congestion of transposed multiples', *Philips Telecommun. Rev.*, 1957, **17**, pp. 145–154

82 LOTZE, A.: 'Traffic variance method for gradings of arbitrary type'. Proceedings of the 4th International Teletraffic Congress, London, 1964, Paper no. 80

83 AITKEN, A.C.: 'Statistical mathematics' (Oliver and Boyd, 1939)

84 BEAR, D.: 'The use of 'pure chance' and 'smoothed' traffic tables in telephone engineering'. Proceedings of the 2nd International Teletraffic Congress, The Hague, 1958

85 WARMAN, J.B., and BEAR, D.: 'Trunking and traffic aspects of a sectionalised telephone exchange system', *Proc. IEE*, 1966, **113**, pp. 1331–1343

86 KUMMERLE, K.: 'An analysis of loss approximations for link systems', preprints of technical papers, 5th International Teletraffic Congress, New York, 1967, pp. 327–336

87 SMITH, N.H.: 'More accurate calculation of overflow traffic from link-trunked crossbar group selectors'. Proceedings of the 3rd International Teletraffic Congress, Paris, 1961, Paper no. 36

88 LOTZE, A.: 'Optimum link systems', preprints of technical papers, 5th International Teletraffic Congress, New York, 1967, pp. 242–261

89 POLLACZEK, F.: 'Uber eine Aufgabe der Wahrscheinlichkeitstheorie', *Math. Zeit.*, 1929–30, **32**, pp. 64–100 and 729–750

90 ELLDIN, A.: 'Further studies in gradings with random hunting', *Ericsson Tech.*, 1957, **13**, pp. 177–257

91 COLEMAN, R.D.: 'Use of a gate to reduce the variance of delays in queues with random service', *Bell Syst. Tech. J.*, 1973, **52**, pp. 1403–1422

92 RIORDAN, J.: 'Delay curves for calls served at random', *ibid.*, 1953, **32**, pp. 100–119

93 WALLSTROM, B.: 'Artificial traffic trials in a two-stage link system using a digital computer', *Ericsson Tech.*, 1958, **14**, pp. 259–289

94 COBHAM, A.: 'Priority assignment in waiting line problems', *J. Oper. Res. Soc. Am.*, 1954, **2**, pp. 70–76, and 1955, **3**, pp. 547

95 COHEN, J.W.: 'Certain delay problems for a full availability trunk group loaded by two traffic sources', *Commun. News*, 1956, **16**, pp. 105–115

96 WAGNER, W.: 'On combined delay and loss systems with nonpre-emptive priority service', preprints of technical papers, 5th International Teletraffic Congress, New York, 1967, pp. 73–84

97 PRATT, C.W.: 'A group of servers dealing with queuing and non-queuing customers'. Proceedings of the 6th International Teletraffic Congress, Munich, 1970, pp. 335/1–8

98 Telephone traffic theory tables and charts (Siemens Aktiengesellschaft, Berlin–Munich, 1970)

99 GOSZTONY, G.: 'Full availability one-way and both-way trunk groups with delay and loss type traffic, finite number of traffic sources and limited queue length'. Proceedings of the 7th International Teletraffic Congress, Stockholm, 1973, pp. 341/1–8

100 BURKE, P.J.: 'The output of a queuing system', *J. Oper. Res. Soc. Am.*, 1956, **4**, pp. 699–704

101 THIERER, M.H.: 'Delay systems with limited accessibility', preprints of technical papers, 5th International Teletraffic Congress, New York, 1967, pp. 203–213

102 COFFMAN, E.G., and KLEINROCK, L.: 'Some feedback queuing models for time-shared systems', *ibid.*, 1967, pp. 288–304

103 Teletraffic Engineering Manual (Standard Elektrik Lorenz AG, Stuttgart, 1966)

104 CLOS, C., and WILKINSON, R.I.: 'Dialling habits of telephone subscribers', *Bell Syst. Tech., J.*, 1952, **31**, pp. 32–67

105 WIKELL, G.: 'On the consideration of waiting times of calls departed from the queue of a queuing system', preprints of technical papers, 5th International Teletraffic Congress, New York, 1967, p. 498 (abstract only – paper presented separately)

106 WIKELL, G.: 'Manual service criteria'. Proceedings of the 6th International Teletraffic Congress, Munich, 1970, pp. 134/1–6

107 CCITT, 1964
(a) 'Measurements of traffic flow'. Recommendation E90, Blue Book, **II**, p. 143 and Q 80, Blue Book, **VI**, p. 119
(b) 'Determination of the number of circuits necessary to carry a given traffic flow'. Recommendation E95, Blue Book, **II**, p. 203 and Q 84, Blue Book, **VI**, p. 125

108 SMITH, N.M.H.: 'Erlang loss tables and other parameters for normally distributed offered traffics'. Proceedings of the 4th International Teletraffic Congress, Paper no. 100

109 HAYWARD, W.S., and WILKINSON, R.I.: 'Human factors in telephone systems and their influence on traffic theory especially with regard to future facilities'. Proceedings of the 6th International Teletraffic Congress, Munich, 1970, pp. 431/1–10

110 WILKINSON, R.I.: 'Non random traffic curves and tables for engineering and administrative purposes' (Traffic Studies Center, Bell Telephone Laboratories, 1970)

111 COHEN, J.W.: 'The generalised Engset formula', *Philips Telecommun. Rev.*, 1957, **18**, 158–170

112 DARTOIS, J.P.: 'Lost call cleared systems with unbalanced traffic sources'. Proceedings of the 6th International Teletraffic Congress, Munich, 1970, pp. 215/1–7

113 HAYWARD, W.S.: 'Traffic engineering and administration of line concentrators'. Proceedings of the 2nd International Teletraffic Congress, The Hague, 1958, Paper no. 23

114 RODRIGUEZ, A., and DARTOIS, J.P.: 'Traffic unbalances in small groups of subscribers'. *Electr. Commun.*, 1968, **43**, pp. 173–180

115 FRANK, H., and FRISCH, L.: 'Communication, transmission and transportation networks' (Addison Wesley, 1971)

116 FRANK, H., and CHOU, W.: 'Topological optimisation of computer networks', *Proc. IEEE*, 1972, **60**, pp. 1385–1397

117 ACKOFF, R.L., and SASIENI, M.W.: 'Fundamentals of operational research' (Wiley, 1968)

118 ANDERBERG, M., FRIED, T., and RUDBERG, A.: 'Optimisation of exchange locations and boundaries in local telephone networks'. Proceedings of the 7th International Teletraffic Congress, Stockholm, 1973, pp. 424/1–7

119 LEIGH, R.B.: 'Standards for the economic provision of direct routes'. Proceedings of the 7th International Teletraffic Congress, Stockholm, 1973, pp. 522/1–5

120 BROCKMEYER, E.: 'The simple overflow problem in the theory of telephone traffic', *Teleteknik*, 1954, **5**, pp. 361–374

121 KATSCHNER, L.: 'Service protection for direct final traffic in DDD-networks', *Nachrichtentech. Z.*, 1974, **27**, pp. 480–484

122 PRATT, C.W.: 'The concept of marginal overflow in alternate routing', *Aust. Telecommun. Res.*, 1967, **1**, pp. 76–82

123 CHAN, W.C., and CHUNG, W.K.: 'Waiting time distribution in computer controlled queuing system', *Proc. IEE*, 1971, **118**, (10), pp. 1378–1382

124 RAPP, Y.: 'Some economic aspects on the long-term planning of telephone networks', *Ericsson Rev.*, 1968, **45**, Pt. 1, pp. 61–71, Pt. 2, pp. 122–136

125 FURNESS, K.P.: 'Time function iteration', *Traffic Eng. & Control.*, 1965, 7, pp. 458–460

126 SINKHORN, R.: 'Diagonal equivalence to matrices with prescribed row and column sums'. *Am. Math. Monthly*, 1967, **74**, pp. 401–405

127 BEAR, D., and SEYMOUR, C.A.: 'A traffic prediction model for a telephone exchange network'. Proceedings of the 7th International Teletraffic Congress, Stockholm, 1973, pp. 536/1–5

128 BACHARACH, M.: 'Biproportional matrices and input–output change' (Cambridge University Press, 1970)

129 WILSON, A.G.: 'A statistical theory of spatial distribution models', *Transp. Res.*, 1967, **1**, pp. 253–269

130 EVANS, A.W.: 'The calibration of trip distribution models with exponential or similar cost functions', *ibid.*, 1971, 5, pp. 15–38

131 WILSON, A.G.: 'Advances and problems in distribution modelling', *ibid.*, 1970, 4, pp. 1–18

132 WILSON, A.G., and HEGGIE, I.G.: Discussion article, in, *Oper. Res. Quarterly*, 1969, 20, pp. 489–496

133 BECKMANN, M.J., and COLUB, T.F.: 'A critique of entropy and gravity in travel forecasting', *in*, 'Traffic flow and transportation'. NEWELL, G.F., (Ed.) (American Elsevier Publishing Co., 1972), pp. 109–117

134 AKIMARU, H., and NISHIMURA, T.: 'The derivatives of Erlang's B-formula.' Review of the Electrical Communication Laboratory, 1963, 11, 428–444

135 ICKLER, S.R., and FLACHS, G.M.: 'A probabilistic trip distribution model with calibration technique', *Transp. Res.*, 1972, 6, pp. 113–117

136 SMYTH, D.G.: 'An adaptive forecasting technique and traffic flow matrix forecasting', *Aust. Telecommun. Res.*, 1969, 3, pp. 38–42

137 BROWN, R.G.: 'Statistical forecasting for inventory control' (McGraw Hill, 1959)

138 PALM, C.: 'Intensitatschwankungen im Fernsprechverkehr', *Ericsoon Tech.*, 1943, 29, pp. 1–189

139 IVERSON, V.B.: 'Analysis of real teletraffic processes based on computerised measurements', *ibid.*, 1973, 29, pp. 3–64

140 TOCHER, K.D.: 'The art of simulation' (English Universities Press, 1963)

141 NAYLOR, J.H., BALINFY, J.L., BURDICK, D.J., and CHU, K.: 'Computer simulation techniques' (Wiley, 1965)

142 Reference tables based on A.K. Erlang's interconnection formula (Siemens & Halske Aktiengesellschaft, 1961)

143 POLLACZEK, F.: 'Problemes stochastiques posés par le phénomène de formation d'une queue d'attente à un guichet et par des phénomènes apparentés', *Memor. Sci. Math.*, Paris, 1957, No. 136

144 NIVERT, K., and VON SCHANTZ, C.: 'Some methods for improving the efficiency of simulation'. Proceedings of the 7th International Teletraffic Congress, Stockholm, 1973, pp. 214/1–5

145 GRANTGES, R.F., and SINOVITZ, N.R.: 'NEASIM: a general purpose computer simulation program for load-loss analysis of multi-stage central office switching networks', *Bell Syst. Tech. J.*, 1964, 43, pp. 965–1004

146 FELLER, W.: 'An introduction to probability theory and its applications' (Wiley, 1968, Vol. 1, 3rd edn. and 1971, Vol. 2, 2nd edn)

147 KHINCHINE, A., Ja.: 'Mathematical methods in the theory of queuing' (Griffin , 1960)

148 HERZOG, U.: 'A general variance theory applied to link systems with alternate routing' preprints of technical papers, 5th International Teletraffic Congress, New York, 1967, pp. 398–406

149 Tables for overflow variance coefficient and loss of gradings and full availability groups (Institut fur Nachrichtenvermittlung und datenverarbeitung der Technischen Hochschule, Stuttgart, 2nd edn., 1966)

150 Tables for variance coefficient *D* and overflow traffic *R* of one stage gradings with limited access. Calculation of secondary route in case of offered overflow traffic (*R*, *D*), *ibid.*, 1965

151 BERRY, L.T.M.: 'An explicit formula for dimensioning links offered overflow traffic', *ATR*, 1974, 8, pp. 13–17

152 KOSTEN, L., MANNING, J.R., and GARWOOD, F.: 'On the accuracy of measurements of probabilities of loss in telephone systems', *J.R. Stat. Soc.*, 1949, B11, pp. 54–67

153 PALM, C.: 'Table of the Erlang loss formula' (L.M. Ericsson, Stockholm, 1954)

154 HOLTZMAN, J.M.: 'Analysis of dependance effects in telephone networks', *Bell Syst. Tech. J.*, 1971, 50, pp. 2647–2662

155 KOSTEN, L.: 'Stochastic theory of service systems' (Pergamon, 1973)

156 BRETSCHNEIDER, G.: 'Repeated calls with limited repetition probability'. 6th International Teletraffic Congress, Munich, 1970, pp. 434/1–5

157 KLEINROCK, L.: 'Queuing systems – Vol. 2' (Wiley, 1976)

Index